A Mile Deep and Black as Pitch

A Mile Deep

and

Black as Pitch

An Oral History of the Franklin and Sterling Hill Mines

by

Carrie Papa

The McDonald & Woodward Publishing Company
Blacksburg, Virginia
2004

iii

The McDonald & Woodward Publishing Company
Blacksburg, Virginia and Granville, Ohio
www.mwpubco.com

A Mile Deep and Black as Pitch: An Oral History of the
Franklin and Sterling Hill Mines

All rights reserved. First printing June 2004
Printed in the United States of America
by McNaughton & Gunn, Inc., Saline, Michigan

10 09 08 07 06 05 04   10 9 8 7 6 5 4 3 2 1

Library of Congress Cataloging-in-Publication Data

A mile deep and black as pitch : an oral history of the Franklin
and Sterling Hill mines / [edited] by Carrie Papa.
    p. cm.
    Includes bibliographical references and index.
    ISBN 0-939923-90-4 (pbk. : alk. paper)
    1. Mines and mineral resources--New Jersey--Sussex County--
History. 2. Sussex County (N.J.)--History. 3. Miners--New Jer-
sey--Franklin (Sussex County)--Biography. 4. Miners--New Jer-
sey--Ogdensburg--Biography. I. Papa, Carrie, 1926-
    TN24.N5M55 2004
    338.4'762234'0974976--dc 22

                                                2004008634

Contents

This book is dedicated
to the memory of
Paul Chandler Moore,

a miner
and my father

Preface

A Mile Deep and Black as Pitch follows my four previous works of oral history, *Bicentennial Voices, Stones and Stories: An Oral History of the Old Monroe Schoolhouse, Farm Women of Sussex County,* and *The Carousel Keepers: An Oral History of American Carousels.* Colleagues and friends in the historical and academic communities in New Jersey urged that I undertake the compilation of an oral history of the Franklin and Sterling Hill mines, located near the neighboring villages of Franklin and Ogdensburg, respectively, in Sussex County, the northernmost county of New Jersey. These mines are known throughout the world for their magnificent and diverse mineral deposits. More than 340 different minerals have been found in these remarkable mines, and thirty species are found at no other place on Earth. For over three hundred years, these rare deposits attracted geologists, mineralogists, and mining experts from all over the world. When the Smithsonian Institution opened a replica mine in 1997 at the National Museum of Natural History in Washington, DC, the exhibit included six tons of minerals from the Sterling Hill Mine. However, after the ore body of the celebrated Franklin Mine was exhausted in 1954 and operations in the Sterling Hill Mine were stopped in 1986, one of the great mining enterprises of the world came to an end.

The idea of conducting an oral history of the Franklin and Sterling Hill mines was personal to me. I had been raised in the small mining town of Franklin, where I lived with my mother and father and three sisters. After graduating from Franklin High

School in 1944, I left the area and did not return, except for brief visits, for thirty-five years. One of those short returns, in 1951, was for the funeral of my father, Paul Chandler Moore. My father had been a miner and a remarkable person. Although suffering with cancer for more than ten years, Dad never was known to complain. After recuperating from five surgeries, he returned to the physically demanding work of being a miner. With frail health, mining must have been extremely difficult for him, yet he cheerfully went off to work each day. Although highly intelligent, Paul Moore had considered himself fortunate to be employed by the New Jersey Zinc Company as a "mucker," the lowest paying job in the mine. During the Great Depression, with a wife and four little girls to feed, he appreciated having any kind of a job.

An appreciation for life, an ability to be happy, to find good in every circumstance, and to remain optimistic in the most trying conditions were outstanding characteristics of this unusual man. Even though he was only a mucker, Dad knew and liked everyone in the mine and in the town. Because he so enjoyed people, he was always a favorite. My father's enthusiasm and appreciation for life were matched only by his rare sense of humor. Through all of those operations, he often joked that he was getting his money's worth out of the health insurance payments. Dad even made his time in the hospital worthwhile. As soon as he was able to get around, he went from patient to patient in the ward to find out if there was anything he could do for them. Within a short time, Dad knew all of the patients, and with his kind smile and optimistic attitude, the entire ward seemed brighter for having him there.

Dad was an avid reader and loved books of every type. He also was quite a linguist and spoke with his mining friends in several of their Slavic languages. Although there was no one with whom he could speak French in the mine, Dad spent many hours studying that language too. My father really was an inspiring man who showed everyone around him how to accept and live each day to the maximum, whatever it held of joy or sadness.

When, after his sixth operation, cancer took his life, his obituary noted only that Paul C. Moore, *a miner*, had died after a long illness. To have the totality of this fine man condensed to the one word, *miner*, seemed to me to diminish his life. To abridge his uncomplaining spirit, his example in living, his endurance and strong character to one word that defined a menial occupation appeared to me to be so insufficient. The word *miner* seemed acutely inadequate to signify all that was his life.

Reading that old obituary some forty years later, however, I reflected that maybe the word *miner* signified more than I had realized earlier. Only recently had I been asked to undertake the oral history project on mining, and I subsequently decided that doing so would give me an opportunity to discover as much as possible about what it had meant to have been a miner in Franklin and Ogdensburg when the mines were in operation. I called the research that followed the Mining Oral History Project; *A Mile Deep and Black as Pitch* is one result of my search for a better understanding of what it meant to have worked in and around the Franklin and Sterling Hill mines.

Thirty-four narrators contributed their recollections to my project, and the pictures they portray constitute this book. Through their commentaries, we gain a heightened sense of the people, the feelings, the places, and the events that made up the Franklin and Sterling Hill mining communities during much of the twentieth century. Through the memories of these narrators, we step back in time to discover not only the hard work and long hours demanded by the mining, but also the satisfaction and pride that the miners took in their work. As we share their experiences, their successes and their failures, their joys and their problems, we honor the lives of these miners and pay tribute to the work they once did in the underground mines of America.

∾

I extend sincere appreciation to the New Jersey Historical Commission for the local history grant that made this Mining Oral History Project possible. I offer my love and gratitude to my husband Albert for his support and patience during the months of interviewing, transcribing, editing, and writing, and to my daughter Jo Anne for all of her recommendations and editing assistance. Also, I thank Alexis Kulick for assistance with the computer.

I extend my deep appreciation to Carol Boone, Lauren Bobier, Judy Moore, and Jerry McDonald at McDonald & Woodward Publishing Company for all of their help.

I particularly want to acknowledge and thank Ann Trofimuk for her suggestions, enthusiasm, and belief in this work, and I appreciate the interest and advice of Pete J. Dunn. For their contributions to the interviews, I thank Sylvia Hadowanetz, Alison Littell McHose, Estelle Mindlin, Julia Novak, Margaret Revay, Joyce Romaine, Helen Shelton, Mrs. Janie Smith, and Irene Zipco.

A Mile Deep and Black as Pitch would not have been possible without the additional photographs and reference material supplied to me by many other individuals and institutions. I am particularly indebted to Melody Bragg, Library Clerk, and Jerry Herndon, Mining Subject Specialist, Library of the National Mine Safety and Health Administration; David Blumenstein, United States Department of Labor; Katharine G. Abraham, commissioner, Bureau of Labor Statistics; Michele Simmons, United States Department of the Interior; Kelly Spinks, United States Postal Service; Michael D. Abbott, Virginia Department of Mines, Minerals and Energy, Commonwealth of Virginia; Michael Hall, Crater of Diamonds State Park; Carol Sheppard, National Mining Association; Jackie Dorr and Nelson Fugate, Mineral Information Institute; Bruce D. Whitehead and W. S. Adamson & Associates for Louis J. Cononelos and Kennecott Utah Copper; Paul Bateman, Gold Institute; Doug Hock, Newmont Mining Company; Robert A. Pond, Frontier-Kemper Constructors, Inc.; John J. Orologio, Society for Mining, Metallurgy, and Exploration; Edward Skipworth, Special Collections and University Archives, Rutgers University Libraries; Sally McGrath, Franklin Historical Society; Richard Hauck, Sterling Hill Mining Museum; John Cianciulli and Ernest Duck, Franklin Mineral Museum;

Stephen Phillips, Buckwheat Pit and Trotter Mineral Dump; Christopher D. Barkley, Windber Coal Heritage Center; Dennis Carollo, Iron Mountain Iron Mine; Steven Ling, Pennsylvania Historical and Museum Commission; Chester Kulesea, Pennsylvania Anthracite Heritage Museum; Leah Greksa, Georgetown Loop Railroad; Val Roy Berryman; Robert and Ann Marshall; and Samuel and Sharon Dunaway.

Finally, I offer a very sincere thank you to each and every one of those thirty-four individuals who so graciously shared their memories and photographs with me: John L. Baum, Clarence Case, Susan Cooper, Alvah Davis, Linda Deck, William C. Dolan, Laura Falcone, Ewald Gerstmann, Wasco Hadowanetz, Richard Hauck, Robert Hauck, John Kolic, Donald L. Kovach, Bernard Kozykowski, Robert E. Littell, William A. May, Robert W. Metsger, Edward S. Mindlin, Steve Misiur, John R. Naisby Jr., Stephen Novak, John L. Pavia, Steve Revay, Harry Romaine, Evelyn Sabo, Steven Sanford, Robert C. Shelton, Thomas G. Sliker, Genevieve Smith, Robert Svecz, Alan Tillison, Ann Trofimuk, James Van Tassel, and Nicholas Zipco.

Introduction

Minerals and the products made from them have shaped the content of human cultures and influenced the course of civilizations. Labels such as Paleolithic, Neolithic, Copper Age, Bronze Age, and Iron Age bear witness to the importance that minerals have played in characterizing the development of human societies and civilizations. Today's global civilization, and all of our lives as individuals, rests solidly and inescapably upon minerals and the innumerable products made from them. Including the obvious metal and glass items through medicines and music to fertilizers and food additives, the list of uses of minerals is virtually endless and the level of complexity at which these minerals can be used is amazing and bewildering.

Some minerals that occur naturally in usable states can occasionally be collected from the surface of the earth, but most must be mined. Mining is an old endeavor, and has been practiced by humans for tens of thousands of years. The ability to find, recognize, and extract minerals has often proven critical to the life and security of civilizations, kingdoms, and other sovereign states, and just as often the mineral wealth of states has invited envy, conquest, and destruction. Much of the global exploration during the past millennium was undertaken with an eye toward locating valuable mineral resources in unknown lands. European expeditions to the Americas and other parts of the world were searching for, among other resources, gold and any other useful minerals that might be found.

The history of mining in the Americas reflects the occupation and development of the continents by humans. Native Americans exploited and utilized mineral resources from the time they first arrived many thousands of years ago. In the United States, from the early Spanish pursuit of gold in the 1500s, to the Dutch explorations for minerals in New Jersey in the 1600s, to the western gold rushes in the 1800s, to the recent discoveries of oil in interior and offshore locations in the 1900s, the search for minerals has been crucial to both stimulating and supporting the expansion and growth of the country. The New Jersey Zinc Company was one of the pioneers in the development of the large-scale mining industry of the country in the decades following the Civil War. After its consolidation with other mining groups in 1897 it was recognized as a leader in mining circles throughout the United States. In addition to its mines in New Jersey, the company had extensive mining operations in Colorado, Illinois, New Mexico, Pennsylvania, and Virginia.

The New Jersey Zinc Company mines in Sussex County, New Jersey, are acknowledged as the "birthplace" of the zinc industry in the United States, and the company town of Franklin was known as the "model mining town of America."[1] Consequently, the Zinc Company and its New Jersey mines with their two thousand hourly workers and company-supported mining communities of Franklin and Ogdensburg, offer an excellent vantage point from which to take a look at the United States' mining industry from the miner's point of view (figures 1, 2).

While each mining company and every community has its own individual history, the stories of the miners of Franklin and Ogdensburg are parts of a larger economic and social landscape of mining in general. The experiences and perspectives of the zinc miners of Sussex County are representative of, and connected to, the coal miners in Appalachia, the laborers in the western copper mines, and the thousands of miners in a multitude of other places. There are more similarities than differences among them. Many miners were immigrants with little formal education. Mining companies employed these men and demanded extremely difficult and dangerous work for very little pay. The social, educational, and health challenges of the miners and their families were common from community to community. Mining companies often set up and operated their towns

Figure 1. In the early 1950s, the aboveground facilities of the New Jersey Zinc Company plant in Franklin, New Jersey, were the economic center of the community. Zinc from Franklin provided basic raw materials for many of the nation's major industries. (Photo: *Zinc Magazine;* Courtesy of Ogdensburg Historical Society)

Figure 2. This photograph shows the facilities of the Sterling Hill plant in Ogdensburg, New Jersey. Sterling Hill and the plant located in Franklin employed more than two-thousand hourly employees. (Photo: *Zinc Magazine;* Courtesy of Ogdensburg Historical Society)

---

to benefit the company itself rather than the workers or their families. Such conditions eventually led to the development of unions for unskilled laborers; the United Mine Workers of America is one such union.

While many books chronicle the history of mining in the United States and several books detail the history of the New Jersey Zinc Company and the incomparable Franklin and Sterling Hill mines, little has been recorded of the day-to-day lives of the miners, especially as it was seen from their perspectives. *A Mile Deep and Black as Pitch* presents that past in the words of those who lived it — the underground miners who tell us about their work and their lives. The Franklin and Sterling Hill workers speak for thousands of other hard-rock miners throughout the United States who faced daily the same kind of difficult, dangerous work. In hearing from miners who experienced the golden age of the New Jersey Zinc Company, we gain first-hand knowledge of the American mining industry as it existed during the first half of the twentieth century and of the miners who made the industry possible.

Historical Background of Franklin and Ogdensburg

Sussex County, New Jersey, lies within the great Appalachian Highlands, which extend from Newfoundland and the Maritime Provinces of Canada southwesterly to Alabama. The Kittatinny Mountains and the Kittatinny Valley occupy a considerable part of Sussex County. The mining areas of Franklin and Ogdensburg lie in the eastern section of the county, in the Kittatinny Valley, near the headwaters of the Wallkill River. The historical narrative of Franklin and Ogdensburg is inexorably tied to the exploitation of the minerals found in the area. This history begins more than 600 million years ago during the Pre-Cambrian period when the great Franklin ore body, which was originally known as "Mine Hill," and the Sterling Hill ore body were formed. According to geologist John L. Baum, the abundance of minerals in this area exists because of an assortment of geological events seldom if ever duplicated elsewhere:

> *A billion years ago sand, silt, and carbonate were deposited on the bottom of a great sea reaching from*

New Jersey nearly to Hudson Bay. With these sediments were submarine lavas and accompanying hot springs. In quiet deep spots, iron, zinc, and manganese accumulated and were incorporated into the thickening sequence. Over 800 million years ago, depth of burial and folding transformed the rocks and their enclosed cores into the layered materials we see today.[2]

Unlike any other mineral deposits known in the world, the Franklin and Sterling Hill ores contain three principal metals: zinc, manganese, and iron. All three occur in a complex and variable mixture, with some veins containing all three minerals and others consisting of only one. The deposits also are unique in being relatively free of other heavy metals such as lead, cadmium, and copper, which usually are found in zinc-bearing ores. The unusual composition of the Franklin and Sterling Hill ores puzzled and frustrated early prospectors and mining experts alike, from the time they were first explored by European colonists until their true identity was recognized in the middle 1800s.

Although Indians had occupied the Sussex County area for thousands of years and they used diverse mineral resources, there is no evidence that they utilized any of the Franklin or Sterling Hill ores extensively. The Lenni-Lenape, or Delaware, Indians occupied the Kittatinny Mountains before and at the time of European arrivals, and they might have led the Dutch, the first Europeans to arrive in Sussex County, to exposures or "mine holes" from which the Indians had extracted ore. The Dutch extensively explored the region, but there is no evidence that they achieved any success at identifying or mining the unusual ores in the Franklin-Ogdensburg area. In *That Ancient Trail*, Amelia Decker describes "The Old Mine Road" which ran through northern New Jersey to the Delaware Water Gap. She also reports that in Amsterdam, in the National Museum of the Netherlands, one can still see specimens of ore taken from those ancient mine holes.[3] Believing, however, that the ore contained too many impurities to be profitable, the Dutch eventually abandoned their claim.

After the defeat of the Dutch by the English in 1644 and the subsequent British occupation of what is now northwestern New Jersey, the land containing the Franklin and Sterling Hill ores passed through a number of owners, some of whom tried to exploit the mineral resources, albeit with little success. While no authentic date of the mining of these ores can be established, extensive iron deposits were discovered and mined successfully in the county during this period. By the end of the eighteenth century, New Jersey had become a leading iron-producing state; the iron from Sussex County mines contributed a large share of the output. Many of the county's communities, including Franklin and Ogdensburg, had grown up around the furnaces and forges of the iron industry. In fact, Franklin was known originally as Franklin Furnace.

The early efforts to exploit the zinc-bearing Franklin and Sterling Hill ores were tied to the Alexander and Ogden families. In the early land divisions, a part of the Franklin district and all of the Sterling Hill area in Ogdensburg had been granted to James Alexander, a prominent Scotsman who had arrived in the colonies in 1715. His son, William Alexander, was born in New York City in 1726. After the death of his father in 1756, William inherited all of the estate, he assumed the ancestral title of Earl of Stirling, and he became popularly known as Lord Stirling. Upon inheriting the Sussex County properties, he attempted to develop the mineral wealth of both the Sterling Hill and Franklin deposits. The "Lord Stirling Pits" were the earliest known workings of the Sterling Hill deposit, and he sent several tons of "red ore" from the Franklin Mine to England for processing as copper ore — an effort that, for obvious reasons, was not successful. The unique nature of the ores and the confusion over the identity of the various minerals continued to mystify everyone and to reward no one.

Robert Ogden, the patriarch of Ogdensburg and the son of Robert Ogden and Hanna Crane, was born in Elizabethtown, New Jersey, in 1716. When Robert was seventeen years old, his father died and left him a substantial estate. After obtaining a liberal education, Robert married Phoebe Hatfield and eventually settled in the upper Wallkill Valley of Sussex County. Among the sixteen children born to Robert and Phoebe Ogden were their four elder sons: Robert III, Matthias, Aaron, and Elias. All four served with honor during the Revolutionary War and later all were prominent in state politics. During the war, Aaron became an aide-de-camp to William Alexander, Lord Stirling, a circumstance which contributed to the close relationship of the Alexander and Ogden families.

During these years in the area's history, the Ogden family gave its name to Ogdensburg. After the death of Lord Stirling in 1783, Elias Ogden came into possession of the Sterling Hill property. The Ogden family also developed a close connection with the Franklin Mine when Rebecca, the daughter of Robert Ogden III, married Dr. Samuel Fowler. Dr. Fowler purchased the mining area in Franklin from Edward Sharpe in 1810, after his father, Joseph Sharpe, had gone bankrupt. Joseph had obtained the Franklin property in 1750 and had spent many years trying unsuccessfully to develop the unique minerals as iron ore.

A few years after his marriage to Rebecca Ogden, Dr. Fowler also acquired the Sterling Hill property. Like Edward Sharpe and Lord Stirling before him, Dr. Fowler invested much time and effort in attempting to exploit the mineral resources of both communities in a commercially-viable way. Although Dr. Fowler developed an experimental process for the smelting of the zinc ore deposits, the complex chemistry of the ore mass made the procedure too complicated and expensive to produce zinc in sufficient quantity for commercial purposes. Nevertheless, Dr. Fowler is said to have once painted his house with zinc obtained from the ores, and he succeeded in having the United States government use zincite from his Franklin ores to prepare brass for the nation's first set of standard weights and measures. Although he failed in making zinc mining a successful business venture, it was through Dr. Fowler's interest and efforts that the true nature of these ores and minerals was brought to the attention of the geologists and mineralogists of Europe and America. As a result of Dr. Fowler's efforts, Dr. Archibald Bruce of New York, a physician, examined the ores and became the first to correctly identify the "red oxide" deposits as zinc ore. An account of Bruce's findings was published in the American Mineralogical Journal in 1814.[4]

In failing health and with his finances exhausted, Dr. Fowler disposed of his mining properties in Franklin and Ogdensburg in 1836, without ever successfully mining the great wealth of zinc that was in his property.

Dr. Fowler's son, Colonel Samuel Fowler, also was well aware of the potential wealth of the zinc-bearing deposits, and by 1848 he successfully regained possession of the Franklin and Sterling Hill properties. However, Colonel Fowler met with no more commercial success in extracting the various ores than did his father. To raise capital to develop both the Sterling Hill and the Franklin properties, Colonel Fowler sold many of the mineral rights. He accepted shares of stock for the purchase price of some of the mineral rights or property that he conveyed to other groups. He even divided the ores according to the minerals they contained, signing over the zinc ore to one mining company and the franklinite to another, while retaining ownership of the surface land. At one time, Colonel Fowler is reported to have held interests in as many as fifteen different companies.

As a result of Colonel Fowler's complicated transactions, the years between 1850 and 1897 were years of litigation. During this period, several mining companies were organized to exploit the mineral deposits of Franklin and Ogdensburg. Among these were the Passaic Mining Company, the Union Exploring and Mining Company, the Fowler Franklinite Company, the National Paint Company, the Consolidated Franklinite Company, the New Jersey Exploring and Mining Company, and many others that either failed or were absorbed into other companies. Colonel Fowler had many large land holdings and any single parcel of land could have had as many as four subdivisions of title — one each for zinc, franklinite, iron, and the surface — which presented significant difficulties for the companies. Numerous complicated legal contests resulted, creating conditions that stifled growth and commercial success.

In his *History of Sussex and Warren Counties* published in

1881, James P. Snell describes Franklin as a place "which possesses no interest," apart from mining. He adds:

> *The spot is not an inviting one, being inhabited principally by miners, whose cottages are small and scattered at various points without regard to symmetry of arrangement.*

Snell closes his entry on Franklin by listing the post office, school building, church, and store as being "suitable for the demands of the mining population."[5]

Near the end of the nineteenth century, however, the potential commercial value of the Franklin and Sterling Hill deposits had been recognized and a new strategy emerged to take advantage of the opportunity. In 1897, the surviving companies consolidated as the New Jersey Zinc Company, a move that brought about the successful commercial development of the Franklin and Sterling Hill mineral deposits and a significant expansion of the Franklin and Ogdensburg communities. Under the auspices of the New Jersey Zinc Company, Franklin and Ogdensburg evolved from rough mining villages to comfortable, model communities while the company itself became a leader in the nation's mining industry. These conditions of growth, vitality, and security continued for nearly six decades, and it is this phase of Franklin and Ogdensburg's history that is remembered and commemorated in the narratives from the Mining Oral History Project that follow.

Methodology

The purpose of the Mining Oral History Project was to chronicle the lives of miners in the setting of the company towns in which they lived, and to examine their lives within the larger context of the times and conditions that affected their day-to-day existence. To obtain a balanced picture of the place of these miners within the general mining industry, I compared information about the lives and conditions of the Franklin and Sterling Hill miners to that of miners in other parts of the country. I examined working conditions, pay, social expectations, and other characteristics that existed in other mining companies and at different mines to see where the New Jersey Zinc Company and its employees fit into the collective picture of mining life in the United States. The New Jersey Historical Commission partially funded this project.

I determined the scope and goals of the project, established areas of inquiry, completed the background research, and selected potential interviewees. I selected participants for the study, in part, according to their (a) potential for imparting information with objectivity, (b) relationship to the subject, (c) ethnicity, (d) and work experience, education, and social and economic status. I sought diversity in ethnicity, work experience, education, and socioeconomic status. Narrators were nominated by local historians and academicians and by Dr. Pete J. Dunn of the Department of Mineral Sciences at the Smithsonian Institution. Author of the extensive work, *Franklin and Sterling Hill, New Jersey: The world's most magnificent mineral deposits,* Dr. Dunn is the foremost authority on the mineralogy, mines, and miners of the Franklin-Sterling Hill area. To provide additional perspectives on the domestic framework of the miners' lives, I also interviewed other employees of the Zinc Company, other residents of Franklin and Ogdensburg, and the wives and children of miners. The interviews explored both the work of the miner and the underlying culture, the intangible way of life that was, and is, part of the region's mining heritage.

Each of the thirty-four narrators was interviewed alone, with follow-up interviews completed if required. The interviews were conducted and the tapes were transcribed in accordance with the professional standards and guidelines of the National Oral History Association. The memories of participants were cross-checked with newspapers, existing records, historiography, and other interviews to substantiate oral statements. I compared interviews to see how well and in what manner they met internal and external tests of corroboration, consistency, and contradictions. In addition to actual facts, perceptions and judgments of the narrators were included to contribute understanding and impart detail to the historical record.

Selecting material for inclusion in *A Mile Deep and Black as Pitch* was a difficult process because so much of the available information was rich and relevant to my goals for the book. Excerpts finally were selected in accordance with how effectively they tied in with the themes to be explored in the book. Although everything could not be included, it all is pertinent to, and part of, the historical record of twentieth century mining in the Franklin and Sterling Hill mining communities. To make this collection of oral history available to other researchers, complete transcripts of the interviews have been deposited with the Special Collections of Alexander Library, Rutgers University, in New Brunswick, New Jersey.

While every effort has been made to insure accuracy, oral history by its very nature is subjective and must be accepted with that understanding. Inevitably, oral history interviews reveal the particular attitudes and opinions of the narrator. His or her interpretation of events will necessarily reflect individual experience and ideology. Therefore, the purpose of *A Mile Deep and Black as Pitch* is to present a cultural history as interpreted by the narrators from their particular perspectives.

Organization

To impose a degree of order on a complex and extensive subject, *A Mile Deep and Black as Pitch* is divided into three parts: The Mines and the Miners, The Model Mining Town of America, and The Legacy and the Future. Each part contains seven chapters. Part One introduces the narrators and describes the work of the miners who made the American mining industry possible. It documents the actual extraction of the ores from the ground, and describes the process of underground mining, from driving raises and drifts to drilling and blasting. This section presents information about the miners' attitudes and complaints, their hours and pay, and their clothing and food. The dangers and other pressures of mining, safety measures, and technological changes are considered from both the miners' and the company's points of view. This section also explores horseplay at work, superstitions, and mineral collecting.

Part Two describes the lives of miners in the setting of the company towns in which they lived. It discloses every day events

— how the miners lived, their families, their health, their problems, and their hopes for the future. The interviews in this section convey the advantages and disadvantages of living in a one-industry town and the benefits and drawbacks of working for the New Jersey Zinc Company. Part Two also includes the basics of daily life, such as housing and finances, and attempts to compare the narrators' experiences with mining concerns in other company towns.

Part Three reviews the enduring legacies of the New Jersey Zinc Company and other mining concerns, the permanent museums, and existing educational opportunities. This portion covers the final days of the Franklin Mine, the tax fight between the Borough of Ogdensburg and the New Jersey Zinc Company with the subsequent abandonment of the Sterling Hill Mine, and the impact that these events had on the local community. This section examines the development of the Franklin Mineral Museum, the Sterling Hill Mining Museum, the mine replica at the Smithsonian Institution, and the Franklin Historical Society's Heritage Museum. Part Three also looks at the interest of different communities in their mining heritage, discusses some other mines that have been turned into museums, and considers the importance of the American mining industry today.

1. Shuster, *Historical Notes of the Iron and Zinc Mining Industry*, 48.

2. Baum, "The Origin of Franklin-Sterling Minerals," *The Picking Table*, February 1971, 6.

3. Decker, *That Ancient Trail: The Old Mine Road*, 3rd ed., 10.

4. "Archibald Bruce," Appleton's Encyclopedia, n.p.

5. Snell, "Franklin Furnace," *History of Sussex and Warren Counties*, 336.

Part One

The Mines and the Miners

As long as civilization as we know it endures, minerals will be there, playing an essential part in our daily lives.

Women in Mining Education Foundation

The Narrators

Could the little group of determined men that started our company in 1848 have foreseen the struggles against men and nature, which lay ahead of them, we are confident they still would have embarked on their risky undertaking. For the results have shown that they were that type of men.

The First One Hundred Years of the New Jersey Zinc Company

In 1897, all of the various mining interests and properties in Franklin and Ogdensburg were consolidated under the New Jersey Zinc Company. This consolidation marked the birth of the first company specifically organized to mine zinc in the United States and a new era in the history of mining in America. The consolidation of individual and small group businesses into larger corporations was a widespread occurrence during the second half of the nineteenth century throughout the United States. By 1900, as a result of abundant natural resources, numerous technical improvements, an extensive network of railroads, and a labor force bolstered by thousands of new immigrants, the United States had become the leading industrial nation in the world. This was a period of tremendous progress in American industry and the beginning of the golden years for the New Jersey Zinc Company.

At the height of its activity, the New Jersey mines owned by the Zinc Company produced 10 percent of the world's zinc and employed more than two thousand people. Those who recorded their memories for the Mining Oral History Project range from

Figure 3. Left: The horse head trademark of the New Jersey Zinc Company dates back to 1786 when it first appeared on a coin of the Colony of New Jersey. The horse head symbolizes speed, strength, and usefulness to commerce. (Image: *Zinc Magazine;* Courtesy of Ogdensburg Historical Society) **Right:** The horse head symbol is now incorporated into the Great Seal of the State of New Jersey.

Robert Shelton, who started working as an errand boy for the New Jersey Zinc Company in 1916, to Robert Metsger, the last superintendent of the Sterling Hill Mine, when it closed in 1986. Miners, geologists, mine captains, employees in the company store, disgruntled workers, and those who admired the New Jersey Zinc Company all tell their stories of day-to-day life and work in Franklin and Ogdensburg during the past three-quarters of a century.

Together, the following thirty-four narrators recount more than a unique history of a special locality and a matchless mining enterprise. Rather, their experiences mirror the events of mining throughout America and their views reflect the thoughts of thousands of other miners whose names will not be found in this or any other history book. In the words of these men and women we find the universal story of mining. Each narrator, with a brief biographical sketch and an excerpt from his or her interview, is introduced here in alphabetical order.

John L. Baum

Curator of the Franklin Mineral Museum, John Baum was born in New York City, attended private schools, and graduated from Harvard University in 1939. John came to Franklin to work for the New Jersey Zinc Company as a geologist and worked for the Zinc Company for thirty-two years before retiring in 1971. Here John recalls the part of his work that involved prospecting for new ore bodies.

> The geologist had a number of jobs. The mines were just one job. The other job he had to do was prospecting, the evaluating of prospects. The biggest area we had for prospecting was the Franklin-Sterling area. Here we had two ore bodies. This was what the Zinc Company was founded on. And there was a possibility that, in addition to the two ore bodies here, there might be another one hiding. So the Geology Department was very busy prospecting for this ore body. We never found it.

Clarence Case

Clarence Case was born and raised in Ogdensburg. After completing grammar school in Ogdensburg and attending Franklin High School for three years, Clarence quit school at seventeen to start working at Ideal Farms. He started working in

Figure 4. Narrator Clarence Case, miner at Sterling Hill. (Photo: Carrie Papa)

the Sterling Hill Mine in September of 1946. Today, he serves as a tour guide at the Sterling Hill Mining Museum. In this excerpt, Clarence affirms the pride many miners feel and tells why he enjoyed working as a miner.

> You see, my great grand dad worked in Thomas Edison's mine. My great, great uncle also worked for the Edison mine. My dad worked in the mill of the Zinc Company. My two brothers worked down in the mine so I felt, "Hey, why not me too?" I didn't want to be the black sheep of the family. I enjoyed every bit of the time that I worked there. I not only made a fairly good living, but I had the great sense of feeling that I brought the raw material up out of the mine. I would come up and I'd watch these automobiles going by and all the items that the Zinc Company material was made into. That made me have a good feeling — to bring up the raw material and having it made into automobiles, glasses, tires, the whole bit. That's something to be proud of.

Susan Cooper

Susan Cooper was named Ogdensburg Elementary School "Teacher of the Year" in 1997. Susan teaches fifth grade and refers to her students as "Mrs. Cooper's Super Fifth Grade." After taking many classes on field trips to the Sterling Hill Mining Museum, Susan was invited to help create a teacher's guide for the museum. Susan's dedication to teaching and caring for her students is evident in the following.

> This year I've brought every grade level down to the mine, except for seventh and eighth and kindergarten, so you're talking about approximately two hundred students. This will continue with probably more next year. Maybe two hundred fifty. I love being able to interact with all of the different grade levels. When I walk down the hall, I have a first grader yelling to me, "Hey, Mrs. Cooper, look at the rock I found yesterday," or whatever. It's really a lot of fun.

Figure 5. Narrator Alvah Davis, hoist engineer at the Franklin Mine. (Photo: Carrie Papa)

Alvah Davis

Alvah Davis was born in 1905 in Franklin, New Jersey. In 1922, he started working for the New Jersey Zinc Company as a machinist's helper for twenty-six cents an hour. This wage was quite good, considering the economic slump that followed the First World War. Eventually, Alvah was trained as a hoist engineer and worked at that job for the remainder of his thirty-five years with the Zinc Company. Here he depicts the work of a hoist engineer.

> You used to lower the men in the mine. There was two engines. One was just for ore alone, and one for just handling men and supplies and timber and so forth. That was the one I was on. You had thirty men's lives in your hands when that cage is going down mornings and coming up. It was a good paying job, but it was a responsibility.

Linda Deck

Linda Deck is employed by the Smithsonian Institution in Washington, DC. As Coordinator of the Exhibit Hall of Geology,

Gems, and Minerals in the Smithsonian's National Museum of Natural History, part of Linda's work involved the planning and installing of the museum's new mine re-creation and exhibit, which opened in the fall of 1997. The exhibit features ore deposits from four different localities, while the mine re-creation itself is based on the Sterling Hill Mine in Ogdensburg, New Jersey. In her interview, Linda discloses why Sterling Hill was chosen for this honor.

> I'm certainly not an expert on Franklin and Sterling Hill, but what I understand is that it is unique. It is unique in the world. They don't understand what made this terrifically diverse assemblage, really enriched in zinc and iron for sure. But, as I understand it, fully one tenth of all mineral species we know from the entire world, one tenth can be found at Sterling Hill, which is fascinating. What an incredible deposit, especially with this fluorescent feature that it has. Just unique.

∾

William C. Dolan

William Dolan was born and raised in Ogdensburg. When he was just sixteen years old, he followed in his father's footsteps and began working for the New Jersey Zinc Company. Here Bill reports on his first job.

> When I was sixteen, I was a water boy. They used to have a spring out on Sterling Hill, and we used to carry water for the different places in the shop. I was sort of a handy boy carrying water or I had to go get oil or something for somebody else. This was about 1919. Then the "Burg" come along and they put water in and the Zinc Company hooked into borough water.

∾

Laura Falcone

Daughter of miner Paul Chandler Moore, Laura (Moore) Falcone grew up in Franklin. She was the only one of Paul and Lucretia's four daughters to be born in a hospital. In the following excerpt, Laura explains why her parents came to live in Franklin.

Well, it was in the Depression, in the '30s. We had been living in a mountainous locality where there was no opportunity for work. And with four children and starvation facing you, you came to where you could possibly get some kind of work. My mother's sister was Mamie Strait and her husband James Strait worked at the Franklin Mine. Somehow he managed to get my father a job there.

Ewald Gerstmann

Ewald Gerstmann's family immigrated from Germany to the United States when he was five years old. Although his father worked in the Franklin Mine, Ewald himself never became a miner. He did, however, develop one of the finest collections of Franklin minerals in the area. On starting his collection, Ewald explains:

It started in '58. If you remember back in them days, people, the miners, they had minerals, but they really didn't know what they had. They just accumulated them. I bought a lot of material because in them days it was cheap.

Figure 6. Narrator William Dolan started working for the New Jersey Zinc Company when he was sixteen years old. (Photo: Carrie Papa)

Wasco Hadowanetz

Local historian Wasco Hadowanetz was a founding member of the Ogdensburg Historical Society. In addition to serving as a trustee of the Sterling Hill Mining Museum, Wasco has served as the borough historian for several years. He is also vice president of the Sussex County Historical Society and serves as chairman of its Historical Preservation Committee. Wasco attributes his interest in local history to his lifelong residency in the area. Wasco feels that the two years he worked for the Zinc Company gave him an appreciation for what his father and other workers went through. He says:

> While I was working there, I did run across my father in the mine a couple of times. I saw the conditions under which they worked. They worked hard. They spent a third of their lives under there. He talked about it because he knew how to mine and I could see he was pleased when he did something well.

Richard Hauck

Richard Hauck, co-founder of the Sterling Hill Mining Museum, has been committed to mining and minerals since he was first introduced to them at age thirteen. Dick and his brother Robert obtained the Sterling Hill Mine in 1989 and developed the site into a mining museum. In the passage below, Dick explains the circumstances that first sparked his interest in minerals.

> [My interest] goes back to when I was a thirteen-year-old boy. In our junior high school system at that time, every Wednesday morning, you'd have club period. You could go to botany, photography, Latin club, Spanish club, many diverse clubs or sports activities. I chose to take study hall. I figured I wouldn't have to take homework home the night before. I could do it that morning, you see. Sitting in the study hall, the vice principal came in and says, "Fellows, poor Mrs. Sherlock, nobody is there in her mineral club. If we don't get at least three or four people up there, she won't have her mineral club this year." So I got volunteered. That was the start of it.

Robert Hauck

Younger brother of Dick and co-founder of the Sterling Hill Mining Museum, Robert Hauck also developed an interest in minerals at an early age. He had been collecting minerals for forty years, but when the opportunity arose to purchase the Sterling Hill Mine, it was not an easy decision for him to make. Purchasing the mine not only meant a total investment of their life savings for both brothers, but it also meant committing themselves and their families to an uncertain future. After trying unsuccessfully to interest the government and others in preserving the property, the Hauck brothers became aware that if Sterling Hill was to be saved, it was up to them to do it. As Bob says:

> At first it was thought the government, the state, would take this over and maybe run it as a park. But it became obvious they weren't going to do that. Somebody had to come in. We had tried so hard to get others to take over. It was felt we should put our money where our mouth was and do it. So we did. It wasn't done as an investment. We went into it knowing it wasn't going to be an investment. We wanted to save it and set it up in a way that it would perpetuate and end up as what it is really. This is what was within our ability to do so we did it. Eventually, this place will outlive us, which is what we want.

John Kolic

John Kolic's interest in minerals developed after a trip to the Grand Canyon, during which he saw various rocks and minerals as he hiked the Bright Angel Trail. When he returned to New Jersey, he took a job with the New Jersey Zinc Company in Sterling Hill with the intention of collecting specimens. John explains:

> Basically, I went there with the idea of hanging on long enough to get some good minerals. I never envisioned that I would like mining, but it turned out that I liked the work. Being a miner was my favorite job.

Donald L. Kovach

Before he invested in a gas station, Donald Kovach's father worked in the Franklin Mine. The summer Don graduated from Franklin High School, he too became a miner. He had won a football scholarship to college and wanted to make some money before going off to school. In this excerpt from his interview, Don explains why he chose to work in the mine rather than in the mills.

> I did work in the mine because it paid more than if you worked "on top." So I asked when they interviewed me if I could go in the mine and they said, "Certainly, if you want to." Most young men did not. They stayed on what they called on top, in the mills. I started work in early June at the Zinc Company and left for college in September. About four months. I really liked it. My whole life I'd heard it talked about over the dinner table, amongst my dad's friends. My father-in-law still talks about it.

Bernard Kozykowski

Bernard Kozykowski's interest in mineralogy was first sparked when his uncle Evan Jones, a miner, gave him mineral specimens from the Franklin Mine. After working at Sterling Hill through the 1970s, Bernie served on the board of the Franklin Mineral Museum for many years and helped to establish the foundation that is responsible for the Sterling Hill Mining Museum. He feels that mining is a unique occupation because:

> You develop a tie. Your ties to the mining industry itself are quickly established. You wouldn't find it in other industries, normally. For instance, we call ourselves hard rock miners, which means you're working in mines where the ground is in rock. Very hard rock and the work is different than in a coal mine, where you're dealing with ground that is relatively what they call soft rock. Coal miners, typically, wouldn't have anything to do with a hard rock mine because hard rock mines tend to ramble on forever and go very deep.

Robert E. Littell

Lifelong Franklin resident Robert Littell has been involved in politics, either on the local or state level, for most of his adult life. Currently a senator for the state of New Jersey, Bob also continues to run his family's appliance business. When asked why, considering his means and status, he decided to remain in his home in Franklin instead of moving to a more upscale community, Bob replies:

> I never had a desire to move anyplace else. I've lived all of my life in Franklin. It was great growing up in Franklin. We had a huge neighborhood of kids. Everybody watched out for everybody else. Made sure they never got in trouble. Alison [Senator Littell's daughter] is in Franklin living in the house my mother and father had on Hospital Road. She just had a baby so that's the fifth generation of our family living in this town. There's a lot of history here.

William A. May

Along with dozens of other Cornish miners, Bill May's grandfather came from England to Franklin to work in the zinc mine. Because of the immigrants' mining experience and skill, mining companies, including the New Jersey Zinc Company, actively recruited them. Bill's father also worked for the Zinc Company. After working two years underground and seeing his best friend killed in a mine accident, Bill's father was able to get a job on top. When Bill went to work for the Zinc Company, his father was insistent that he remain aboveground. Bill was able to get a job in the company store. In this section of his interview, Bill describes the townspeople's perception of the company store.

> A lot of people figured that the Zinc Company store was for the upper crust in town, the salaried men or whatever. Actually, it wasn't. It was open to anybody that wanted to come in there. But I think some people thought, "That's not for me to go shopping in there. Lord and Taylor or something when I can go to K-Mart." But everything they handled was of a better grade than you could buy in most of the stores.

Like shirts. They had Van Heusen, Arrow. Neckties were Botany. Socks were Botany. We had women's clothes. I knew different women went in there because of the stockings we sold. The lingerie was better than you could get around. Of course, the price had to be there too.

Robert W. Metsger

Robert Metsger came to work for the Zinc Company as a geologist in 1949. Most of his forty working years were at Sterling Hill, and he was still there when the Zinc Company left in 1989. As the last appointed superintendent, it was his responsibility to close the mine. Here Bob comments on his feelings about the closing.

I hated to see it close. Sure. As far as the pay is concerned, they didn't give the highest pay in the world, but as far as a company to work for, it was great. I really enjoyed working for the Zinc Company, but there were a lot of places that paid better. But Franklin wouldn't even be there if it weren't for the Zinc Company.

Figure 7. Narrator William May descended from a long line of Cornish miners. (Photo: Carrie Papa)

Edward S. Mindlin

Edward Mindlin was born in Franklin, and except for the time he served with the navy during World War II, he lived in the area all of his life. Although he did experience discrimination because of being Jewish, Eddie was and is proud of his heritage. As he explains:

> My father came to America in 1902 from Russia. Postov. My father came to Franklin in 1903, permanent, to live. He was the first Jew in Franklin. He came to Franklin because of the miners. There was a lot of Pollacks, Russians. There was Hungarians. My father didn't speak Hungarian, but they spoke Ersatz. Slovaks. He was a peddler on a horse and wagon before he had a store. My father was the first Jew and we're the last. Wouldn't it be nice if we stayed here until 2003 and that would make a hundred years of Mindlins living in Franklin?

Steve Misiur

Steve Misiur's interest in minerals dates back to his childhood in Elizabeth, New Jersey. Steve's curiosity about the natural world around him was heightened by the fact that his grandfather had worked in the Pennsylvania coal mines. Steve lived in a Polish section of town with a little clay patch right in front of his house. Finding a quartz crystal in this clay patch led to his becoming a "Pebble Pup" at nine years old, and to an interest in minerals that became a lifelong enthusiasm. After doing a great deal of volunteer work at the Sterling Hill Mining Museum, today Steve serves as the museum's curator. On entering the mine adit [mine entrance] he recalls:

> [I felt] quite at home here. My very first entry into the mine itself was July 4th of 1989. Just a couple weeks after Dick and Bob [Hauck] bought it. When I walked into that adit something clicked in my head. I felt like something said to me that I came home. It was a very odd feeling when I walked into that adit for the first time. I'd never been into the mine itself and something said I was at home.

John R. Naisby, Jr.

John Naisby graduated from the University of Delaware in 1934 with degrees in mechanical engineering and electrical engineering. Even though the country was in the midst of the Depression, he was offered two jobs: a teaching position at Cornell University and a job with the New Jersey Zinc Company. John chose the Zinc Company job because:

> I thought I was more suited to the kind of work that I'd be doing. They were interested in an engineer. I got employed as an engineer, but I was here for a year before I even knew where the engineering department was. Then they sent me down to Ogdensburg because they were doing some renovating in the mill down there. I went down there and I worked there for the duration of my thirty-one years.

Stephen Novak

Stephen Novak is a Franklin native who went to work for the New Jersey Zinc Company after graduating from high school in 1930. Steve worked for the Zinc Company for ten years in various jobs including the real estate gang, the time office, and the millwright gang. At the time that Steve was working aboveground, his father was working underground as a repairman in the mine. During World War II, Steve was in the service flying with a bomber group. On Thanksgiving Day, November 24, 1943, he was notified that his father had been killed in the mine. He explains:

> I was flying in a bomber group and after so many hours, you can go home. After I got the information that my father was killed, I told them, I didn't want to go. I had no reason to go home. I had nobody to go home to. They told me to go home. So I went home. I got home Christmas Eve. I was bitter then. Christmas Eve, I'm sorry to say, I drank about a half a bottle of whisky and got drunk. When my father was killed, he used to work nights as a repairman, but this particular time, he worked days. He worked on what they call the head frame, where the ore cars were bringing up ore from the mine and dumping it into the coarse crusher. He was hit by one of

those cars coming down. I asked superintendent McCann how bad was my father banged up? He told me there was hardly a mark on him. I couldn't believe that so I went to Jack Ramsey, the funeral director, I said, "How bad was my father hurt, Jack? I want to know the truth." He said, "If they didn't tell me it was your father, I would never have known it."

John L. Pavia

John Pavia came to the United States from Chile in 1926. He heard of employment opportunities at the Sterling Hill Mine through friends, and started working for the New Jersey Zinc Company when he was nineteen years old. He remained with the company for fifty years, until he retired in 1976. Here John describes how he was able to get married, buy a house, and bring up eleven children, seven boys and four girls, during the Depression.

> You work. Eat. Sleep. And work. I get up six o'clock in the morning. Seven o'clock we're down in the mine already. Then I get out half past three. Then I go to work over in Sparta. Come home sometimes eleven o'clock. Fourteen years, I worked in two places. Was it a good life? That was life. I'm still alive!

Figure 8. Narrator John Pavia worked for the New Jersey Zinc Company for fifty years. (Photo: Carrie Papa)

Steve Revay

Steve Revay was born in Franklin in 1907. Just as he was to begin high school in Newton, he came down with typhoid fever and lost a year of school. Steve never went back to school, but instead started working for the Zinc Company at fifteen years old. Child labor in the early years of the twentieth century was common in the mining industry. State laws against child labor generally were not enforced, and parents, as Steve points out, needed the income.

> I never went back to school because my mother and father needed the money. The Zinc Company was a good place to work, but they never paid no wages. Their wage scale was very low. Thirty-one cents an hour.

~

Harry Romaine

Although he was born in Hamburg, New Jersey, Harry's family moved to Franklin while he was very young. After graduating from eighth grade in the Franklin school system, Harry got a job in a grocery store. Later in life, he went to work for the Zinc Company. He explains:

> I decided I had enough schooling. I guess it was crazy at that time, but I left. Of course, I wanted to go to work and make some money. So I worked in the grocery store. The first supermarket in Franklin. Harry Fishgrund. Self-service. The first self-service supermarket in the county. I worked there for quite a few years. Then, when I was twenty-one, I went to work for the New Jersey Zinc Company in Ogdensburg. There was a little more money.

~

Evelyn Sabo

Evelyn (Moore) Sabo was born and raised in Franklin, and was the youngest daughter of miner Paul Moore. She married Joe, an employee of the New Jersey Zinc Company, but because Joe's father had been killed in a mine accident, his mother was determined that her son not go underground. Evelyn recalls:

> Joe's mother, when they took him and gave him his tour of the mine and everything, she went to the Zinc Company and she talked to one of the big wheels and she said, "Don't you dare put my son in the mine. I lost a husband in the mine and I'm not about to have my son work down there." She said, "If you want him to work for you, you're going to give him a job aboveground." That's when they gave him a job in maintenance.

Steven Sanford

After studying geology in his youth, Steven decided not to pursue a degree, but to remain close to his interest in minerals by working in the Sterling Hill Mine. Currently the assistant manager at the Franklin Mineral Museum, Steve recalls that drug use in the mine became a safety problem during the years of 1970 to 1975. He remembers that anyone caught using drugs while working was fired immediately, as he relates in the following.

> In the seventies when I worked there, just past the psychedelic sixties, there was some drug use underground. Some guys were caught using dope and they left hurriedly.

Robert C. Shelton

Although employed by the New Jersey Zinc Company as a young man, Robert Shelton is best known in Franklin through his employment with Franklin's first bank, the Sussex County Trust Company. Bob started with the bank as a credit manager in 1926. He served with nine different bank presidents and eventually became a director of the bank himself. Here Bob gives a short lesson on the banking business.

> Do you know what the first word is they teach you when you go into the banking business? NO! That's what people think. Do you realize when I turn down your application I'm not making a cent. In fact, you've taken two hours of my time which we're not charging you for. How to say YES! That's the job.

Figure 9. Narrator Thomas G. Sliker was the underground mine superintendent at Sterling Hill for twenty years. (Photo: Carrie Papa)

Thomas G. Sliker

Thomas Sliker started working for the New Jersey Zinc Company in 1941, and remained there until the Sterling Hill Mine ceased operations in 1986. Tom started out as a mucker and closed his thirty-seven year mining career as underground mine superintendent. Tom's enthusiasm for mining and his leadership qualities are evident in everything he says about his work.

> I never had any idea that I was going to be a miner, but when I went there I loved it. I loved it and I learned everything that you could learn about mining. It was dangerous and very hard work, but it was very interesting. I had a lot of experience for thirty-seven years. It was a good life. When I quit at the age of seventy, I had run the mine for twenty years.

Genevieve Smith

Genevieve Smith was born in the Franklin Hospital on Christmas Day, 1920. After graduating from Franklin High

School, Genevieve trained at Philadelphia General Hospital and became a registered nurse. She served as an army nurse during World War II, but then returned to Franklin. Genevieve, one of the few women to be employed by the Zinc Company, worked at the Franklin Hospital, where she spent many of her forty-two years of active nursing. Both Genevieve's father and grandfather had been employed by the Zinc Company, her father having been a general chauffeur for the company and drove the company's first ambulance, the first ambulance in the county. About the hospital and its staff, Genevieve says:

> [The staff was] excellent. Excellent. It was the first hospital in the county. It was well equipped for the work we did. I really can't praise the hospital enough.

Robert Svecz

Robert Svecz was born and raised in Franklin. Both of his grandfathers worked in the mine, as did his father before joining the military in World War II. Having heard about Franklin minerals all of his life, Bob chose to study geology in college. During his summer vacations, he worked as a miner at the Sterling Hill Mine because:

> There was always a kind of fascination about it because everybody talked about the mine, and, of course, the minerals. The fascination about the minerals that were here. It became apparent that this was kind of unique. I wanted to see what it was like. I wanted to see where my grandparents had worked, where my father had worked, what it was like.

Alan Tillison

Alan Tillison grew up in Hamburg, a village two miles north of Franklin. He graduated from Hamburg High School in 1933, but finding a job at that time was not easy. Alan reports:

> That was during the Depression. 1933. You could hardly find a job anywhere. I spent almost two years in the Civilian

Conservation Corps at High Point. Then later I worked on a farm. Then, after we were married, then I got the job as a miner. I went down in the mine for the New Jersey Zinc Company. When you first started out, you were what they called a mucker. You used a shovel! [laughter] It was hard work, but it was good because you had a job! There were a lot of people that didn't have jobs at that time.

Ann Trofimuk

Ann's father and brother, born in Russia, both worked in the Franklin Mine. Her brother Nicholas was a teenager when he arrived in America and he knew no English. Because of this, Nick was put in a special class in school and the other children made fun of him. This experience discouraged Nick from continuing his education. He went to work in the mine as soon as possible. In spite of his lack of formal schooling, Nick became an expert on Franklin minerals. As Ann observes:

> He was very intelligent. If you get into minerals, you will find that he is one of the best sources of minerals and data on minerals. He opened his mind to everything that was there. He was frustrated that he couldn't go on with his education, but he completed it within the mine itself. Part of Nick's collection he sold to Harvard University.

Figure 10. Narrator Alan Tillison was happy to get a job at the Franklin Mine during the Depression. (Photo: Carrie Papa)

James VanTassel

James Van Tassel is a lifelong resident of Franklin. He graduated from Franklin High School, and now works for the Borough of Franklin. Jimmy was the first president of the Franklin Historical Society. In 1996, the society opened the Franklin Heritage Museum in what had once been part of the Zinc Company offices on Main Street. About their new museum and the future of Franklin, Jimmy declares:

> We're very proud of this. The museum is like a home base. It's all been done in less than a year. It's totally remarkable, and everything for the future is like, "Go right ahead." We hopefully have a nice future ahead.

Nicholas Zipco

Nicholas spent some of his childhood in Pennsylvania, where his father worked in a coal mine. When a visitor from Franklin gave Nick two unusual specimens, he developed a passion for mineral collecting and went to work in the Franklin Mine. Today, Nick has one of the finest private collections of Franklin minerals. In the following selection, Nick explains how he learned to distinguish the different minerals.

> I learned by hand. I read the book a lot. Palache's book. Most of these other books are copied out of there, just changed a little bit. Palache describes all the different minerals in Franklin. Local minerals. That's the best book, really.

The memories of these narrators, captured here in their own words, do not only allow glimpses into life and work in the mining communities of Franklin and Ogdensburg. Rather, they chronicle the evolution of company-owned mining towns over the years and across the country. As we absorb the narratives of hardship and sorrow and joy and success, we are reading a story of our country's past.

They Came to Work

Between 1820 and 1930, thirty-eight million Europeans migrated to the United States in what is certainly one of the greatest migrations in history.

Almanac for Americans

During the later years of the nineteenth century and the early years of the twentieth century, the industrial economy of the United States expanded at an unprecedented rate. The widespread increase in factories and manufacturing, made possible in great part by the mining industry, substantially increased the demand for cheap labor. Since the domestic labor supply could not meet the demand, immigrant labor was utilized and immigration was basically unrestricted. More than a million immigrants came into the United States annually during several of the years between 1904 and 1914.

Many of these newcomers found work in either the mills or the mines of America. In 1870, the population of Calumet township in Michigan was 3,182, with 2,051 of that number being foreign born.[1] In Michigan, the majority of immigrants came from the Scandinavian countries, England, and Canada, whereas in Pennsylvania and New Jersey, most immigrants were from the Eastern European countries. The 1922 report of the United States Coal Commission determined that most of the foreign-born laborers who became miners were from England, Austria, Hungary, Italy, Poland, Russia, France, "Slovakland," Yugoslavia, and Germany. The report noted that union meetings in Appalachian coal camps were conducted in three languages. This also was true in Michigan where the Calumet & Hecla Mining Company printed notices in three languages in addition to English.[2]

Among the thousands of immigrants who came to America from Russia were the parents and brother of Ann Trofimuk. Ann's father had come to the New Jersey zinc mines from the coal mines in Pennsylvania, as had several other miners. Depleted mines, bad working conditions, accidents, and strikes were the reasons miners moved from one mine to another. Here Ann observes the social aspects of the immigrant experience and speaks of her father with pride.

> My father was here in this country before the First World War. He was a young man and I'm not too familiar with the circumstances. I guess he heard that America was a land of opportunities. My brother Nick was born and a babe in arms when my father left [Russia] to come to the United States. He enlisted in the US Air Force. All the time he served in the military, Mom and Nick were in Russia during the war. The war ended. He came back to the United States as a member of the military, and he brought my mother and brother over from Russia after that. They were not in steerage because my mother was considered a citizen. My father had applied — as a member of the military after World War I — for citizenship for my mother and brother. As a result, they were citizens. The minute Nick put his foot on American soil he was a citizen. When my brother came to this country, I think he may have been sixteen years old. According to what my brother said, they came on a very nice ship. Were treated very nicely. They came to Hoboken. Then they came by train to Franklin, from Hoboken, and settled in Franklin with my father.

> My father had worked in the mines in Pennsylvania, and, apparently, had heard that there was work available at the Zinc Company during hard times, when work wasn't available elsewhere. So he was working here in Franklin when he brought my mother and brother over from Russia. As somebody who was born to Ukrainian/Russian parentage, we had all the Russian dishes, the borschs, the cabbage soups, the cabbage rolls. My mother had a garden. In the summertime, she grew the vegetables and she canned. I think you'll find that somewhere along the way as the different mothers and ladies and wives in Franklin got to know each other, the cuisine drifted between nationalities.

One lady would tell another lady, "Well, in my country, we made this kind of thing this way." I think there was some flowing back and forth of the cuisine. There was quite a bit of interaction between these various ethnic groups. Definitely. It was like a hodgepodge. Like a small United Nations.

My brother could speak Hungarian quite fluently. He learned it from working with Hungarian men in the mine. I myself picked up a few words of Hungarian from having Hungarian families in the neighborhood. You remember Mrs. Chet? I'd always say that she taught me how to call pigeons in Hungarian. Tyouboo, Tyouboo, Tyouboo. [laughter]

Each nationality — not everyone — but the Russians had their own society. The Slovaks had their own society. The Hungarians had their society. They met once a month. Also, I think they had insurance, which they paid through the group. Then they had their little get-togethers like picnics at Kovach's Grove or dances at Moxie's Tavern, which is now Sullivan's. Or else in the church Lyceum in Franklin. When they had these social get-togethers, if you wanted to go you were welcome to go. There was no entrance fee to my knowledge. They sold hot dogs and sauerkraut, this kind of thing. A perfect example of how things were a mixture and that people got along is one of my fondest memories — Meyer Rosen and his wife Lena. Meyer Rosen had the bakery in Franklin and they were Jewish. I don't know but I assume they also went to the Hungarian dances. The dances, we called them balls. Meyer Rosen and Lena would come to the Russian balls and Lena would come in an evening gown. I don't know at that point whether she was the only one that had evening gowns in Franklin. They would dance and they would request that the orchestra play a rumba and they would get out there and do the rumba. Then you would have the orchestra play the Russian Cossack. I cannot for the life of me understand how some of these then young women in heels and stockings would do the Russian Cossack where they got down and flung their legs out. You think of the Russian groups doing it, the men. But here they are the ladies in stockings and the urban dresses.

Figure 11. The Hungarian Band was organized in 1918. (Photo: Courtesy of Franklin Historical Society)

Bob Svecz, who worked summers at Sterling Hill while he was going to college, reported that both of his grandfathers worked in the mine. They both had come from Eastern Europe, and as with thousands of other immigrants, they had left their homes in the "Old Country" for economic reasons.

> On my mother's side of the family, the Lutes, they came from the Ukraine. My father's side, the Svecz, were Slovak. They came to America for the same reason that everybody came before World War I. They came for the promise of a better life. I wanted to work in the mine because I wanted to see where my grandparents had worked, where my father had worked, what it was like. I kind of enjoyed it. I worked during the summer. But, by the end of summer, I was ready to go back to school. But it wasn't a bad job.

Wasco Hadowanetz is yet another first generation American whose mother and father emigrated from the Ukraine in Russia. Like the Trofimuks, they too had relocated from Pennsylvania to New Jersey. Although they were from neighboring villages in Russia, Wasco's parents did not date or marry until both were living in the United States. Wasco recalls:

They Came to Work

They came from the Ukraine, just before the First World War, 1914. They both came from there. Apparently, they were in neighboring towns because they used to take care of the cattle. They said they remembered each other from there, but they weren't married until over here. My father went to the coal mines [in Pennsylvania] because my uncle, my mother's brother, was working there, and I imagine he was staying with him. Then he met my mother who was working in the Passaic silk mills, so that's how they got together. [My wife] Sylvia and I have gone up there. In fact, we found the church they were married in. Dixonville, Pennsylvania. I imagine when things were going downhill in the coal mines, he sought work at the zinc mines. Around 1927, he moved to Ogdensburg and started in the mine with the New Jersey Zinc Company.

I was born in 1930. We're a family of ten — eight boys and two girls. I was the seventh son, and he waited until the seventh son to name a junior. There's some importance to that. Something to do with the seventh son of a seventh son. My father is a seventh son. So he waited for the seventh son to name him junior. I was born in 1930 and by that time we had seven in the family.

We spoke Ukrainian at home. They learned enough English to get along, but mostly they spoke their own languages. My parents, I didn't even know if they were speaking Russian, Ukrainian, or Polish sometimes. They even knew a little Hungarian from their experiences overseas and then they picked up some more from the neighbors. Hungarians on one side, Mexican family on the other. You had to learn some of the language. There was a sense of community, of neighborhood. I remember sitting on the front porch and Tony Delatorre would bring his guitar over. Matt Smith, who was a Czechoslovakian, I think, would sing. And someone would bring a mouth organ. We'd sit on the front porch and have a band. Down the street I remember Rosie Ramirez. What a voice she had. She would sing and the whole community would hear it and listen to her singing. She would sing some Mexican songs. I remember next door, they used to make tamales from scratch. We also knew a Mrs. Warden who used to bake biscuits and the kids would be down by her porch just as she would take them out of the oven.

Steve Revay, another first generation American, was born in Franklin in 1907. His parents came to Franklin when it was still known as Franklin Furnace. Here Steve explains that the school changed the spelling of his name and comments on the different ethnic groups.

I was born down by the pond. There was a place there that used to be a furnace before the Zinc Company came here. They had a plant down there that made stoves. Franklin Furnace, that's where it got its name. That's where my father lived before I was born. The name Revay is a mixture of Hungarian and Czechoslovakian. It's a well-known name in Czechoslovakia and in Hungary. High class, you know. It was spelled with two capitals, capital R-e-capital-V-double-i-i. When he came over here, they wrote it out as R-e-v-a-i. They called it REVAI. The school changed it to R-e-v-a-y. It had been a double *i* on the end.

The people who worked in the mine were the Pollocks, the Slovaks, the Hungarians, the Russians. Mexicans. If they needed men, they'd go to Mexico and get them. Most of the good jobs went to the English. A lot of English were bosses. The captain in the mine was an Englishman. The Irish had good jobs too. The McEntees, the Flynns, Quinns. But like I said, the Hungarians, Pollocks, the Russians, Slovaks, they worked in the mine. They were the miners because they didn't have no education or anything else. They just put them in the mine and told them what to do and that's it.

Clarence Case, a miner at Sterling Hill for a number of years, agrees with Steve Revay that the bosses were often English. Clarence also feels that an education was not necessary for mining.

You don't need any skills at all. All you have to know is how to use a shovel and have a little brawn. When I first started working there, they gave me a shovel. Clean track. Well, this is fine. Then I worked in timber. Then I was working on the motor as a helper, tramming. Then I was put in a stope in a raise along with experienced runners, and that is where I spent most of my time in the mine, in the raise. Machine runner, that is the man that operates the machine. He also has a helper. I was a helper for awhile until I got graduated into a machine runner. Then they gave me a helper and we got along just fine. These people I worked with were Chileans, Mexicans. Those people were extremely hard working people. Absolutely. They were good workers. But you don't need any skills at all.

Contrary to Clarence's belief that mining did not require skilled labor, a number of the well-educated salaried employees very much admired the miners in spite of their lack of formal education. Steve Novak, whose father was killed in a mining accident in Franklin, reports on his conversation with one of the Zinc Company engineers.

These fellows, Russians, Hungarians, Slovaks, had no formal education. Those fellows that came from Europe maybe only had as high as a sixth grade education. My father only went to the sixth grade. I spoke to an engineer, Steve McPartland, from Princeton. A lot of the engineers, after they get out of college, they work in the mine to get practical experience. This Steve McPartland, I got to talking with him one time and he said, "What you learn from these miners, you'll never get out of books." They could look at a body of ore and tell which way it's going to go when they blast. Art Watt was a very good fellow. He had a lot of faith in these men that worked in the mine. A lot of the English were the bosses and they oversee'd the others. But, this Art Watt, he had a lot of respect for the miners because they knew what they were doing. There was always a mixture of Hungarian, Slovak. All mixtures. People got along quite well. Bosses were very friendly and considerate. Art Watt, he was very good with the fellows.

In the following, John Baum emphasizes his admiration for the skilled miners with whom he worked.

These miners, of course, are accomplished. They're artists in their own way in doing this kind of work. Each one had many jobs. They were, in a way, engineers with the building of timbered frames and what not to hold the ceiling up. Their lives depended on their doing it right, keeping those timbers from falling back down again. It was all done with wedges and props and things and they knew just how to do it. Every time they took out another ten or fifteen feet of ore, they had to put up another set of timbers. They had a boss. He was called a shift boss. He had a bunch of miners under him. We had the Cousin Jacks. They were from Cornwall. Cornwall miners, England. They came from mining operations in the Old Country, many of them did. They had a contact here because they had cousins and whatnot that worked here in the mine. Favorite name over there is John or Jack and so a miner in the very old days would go to the boss and say, "Do you mind if I send for my Cousin Jack?" And this was the case throughout the United States. And they called them Cousin Jacks, that's all there was to it.

And, of course, we had the Hungarians. We had a great many other Eastern Europeans. And Mexicans. And Russians. The mill was especially a favorite working place for Hungarians. The Cousin Jacks — I don't want to be too class conscious here — but they were preferred by the company for the shift bosses. Cornwall miners were the finest miners there were because of their skills. Because they were English. Because they had the characteristics of the English people, especially the Cornwall people. Ingenuity. Careful workers. Ability to work with others. And so on and so forth. To me, they were stubborn as hell. They were stubborn. They knew just what they wanted and would do what they wanted and they did it their way. They were doing right as far as mining was concerned. They were good miners. Very good. World's best. They were brought over here all through the West. They were brought here from Michigan because they knew the mining method, which is called top slice method. A number of the bosses came here for that.

Bill May, a descendant of Cornish miners, worked in the Zinc Company store in Franklin and reports that not only his father, but also both his paternal and maternal grandfathers worked for the Zinc Company. His mother's Cornish father had come to Franklin from the copper mines in Michigan. From 1913 to 1914, a bitter labor struggle had taken place between labor and management in mining companies in the Upper Peninsula, such as the one between the giant Calumet & Hecla Mining Company and the Western Federation of Miners. When the miners went on strike in 1913, nearly sixteen thousand employees were affected. By the end of 1913, about twenty-five hundred miners had left the area to look for work in other mines. Here Bill describes his grandfather's move from Michigan and his friendships with immigrants.

> My father came from England with his parents and his brother. I think it was right after World War I. When he came here, he was only twelve years old. They settled on Church Street. My grandfather worked for the Zinc Company. My father worked for the Zinc Company for thirty-five years. He was a millwright. They did all of the repairs, when they had something break down. There were so many people that came from England, from the Cornwall area that were miners over there. That's why this was a big attraction for them. Most of the foremen in the mine were Cornish. They had the experience. Most of them were miners from England. Also, my mother's side. Her father, Bill Jones, was a mining engineer from the copper country in the Upper Peninsula, Michigan. There was a lot of people from around here from there too.
>
> They had a cross section of Europe working there. In fact, they used to bring Mexican labor up from Mexico in cattle cars years ago. They had their own development in Ogdensburg and they had it in Franklin too. Little Mexico. They had big boarding houses. There were a couple — there's still two buildings on Buckwheat Road here that have been fixed up into nice big homes. I went to school with a boy named David Garcia. He was Mexican. In fact, there was a girl I went to school with, her father was a Peruvian. Peruvian was tin mining. Peruvians came up here too. One was Pavia. Johnny Pavia, I knew him from sports.

Actually, the Johnny Pavia whom Bill May remembered as Peruvian was the son of John Pavia, who came from Chile. John Pavia's godmother, who was living in New York at the time, helped him get to America in 1926. Then, through an Italian friend, John learned of work in Ogdensburg with the New Jersey Zinc Company. When he arrived in Ogdensburg, he did not know if he would be working in the mine or outside, but times were tough so John told the company:

> I'll work any place as long as I get some work. They gave me work. They needed work in the mine so I worked there. All kinds of jobs. Mucker. Miner. Everything was pick and shovel. We had these small cars. You fill them up and you had to push the car, all handwork. It was quite heavy. We used to push the car quite a ways. Maybe two hundred feet. Then we dumped it into the grizzly. Sometimes we loaded twenty-four cars in a cut. But in the mine, everything is hard work. I was a top worker in there. Any job they wanted to get it done, they got me. Fifty years, I worked with Hungarians, Slovak, Russians and different nationalities. This mine was like a League of Nations. [laughter]

Figure 12. This miner is pushing a car loaded with copper ore at a Calumet & Hecla mine in Michigan. As narrator John Pavia says, "We had these small cars. You fill them up and you had to push the car, all hand work." (Photo: Courtesy of Val Roy Berryman)

Whether an immigrant to the United States, a first generation American, or a person whose ancestry could be traced to colonial America, everyone who participated in the Mining Oral History Project shows a distinct sense of pride in the New Jersey Zinc Company and the mining towns of Franklin and Ogdensburg. Although there is an honest criticism of things they felt were wrong with the towns or with the Zinc Company, the narrators still retain a strong sense of respect for the company, the mine, the miners, and the lives they lived in a unique mining area.

Participants in other mining oral histories, such as those quoted in *Appalachian Coal Mining Memories*, have spoken with similar regard about their own operations. Miners in Pennsylvania and Michigan displayed the same sense of community pride. Laura Falcone, daughter of miner Paul Moore, speaks for most mining families when she says:

> The whole town was a mining town and no matter what position you might have had, you were still a miner. One of my old school teachers, Mrs. Higgins, told the class about zinc mining. That it was one of the most wonderful mines in the world. That Franklin was the model mining town of America. Community pride and a sense of belonging were instilled in us. And it was a beautiful town.

1. Thurner, *Calumet Copper and People*, 44
2. US President's Commission on Coal, *The American Coal Miner*, 20.

Miners, Muckers, and Harvard Graduates

From 1897 on, the operation of the Franklin Mine was highly successful. Under the direction of the Zinc Company and with the utilization of the most modern technology, the Franklin Mine became one of the foremost mining operations in the world.

Franklin Borough Golden Jubilee

There were several differences between the Franklin Mine and the Sterling Hill Mine. While they both contained many of the same ores, the actual ore veins were dissimilar. Also, Sterling Hill was a considerably deeper mine than Franklin. In fact, two buildings, each taller than the Empire State Building, could be stacked one on top of the other in the depths of Sterling Hill. While deep, Franklin and Sterling Hill were nowhere near the depths of many of the Pennsylvania coal mines, where some extended more than three thousand feet below the surface, or of the copper mines in Calumet, Michigan, where several were over a mile deep. The *Guinness Book of World Records* lists the Western Deep Levels Mine at Carletonville, South Africa as the deepest mine in the world. Going down an extreme depth of 11,752 feet, the rock in the lower levels of this mine reaches temperatures of 131 degrees Fahrenheit and necessitates refrigerated ventilation. Also, pressures on the rock at that level are a continuous hazard.[1]

In an article in *The Picking Table*, Clarence Haight makes the geometry of mining understandable to a lay audience by comparing the mine to a city office building.[2] Haight, former superintendent of Franklin Mine, writes that the building would be about a mile long, five hundred feet wide, and a thousand feet high. Of course, this building is not on the surface, but rather a thousand feet deep in the earth. The floors of the building are the "levels" of the mine which, instead of being the usual eight or ten feet apart vertically, as in an office building, are from fifty to one hundred feet apart vertically. The office elevators are the mine's "shafts," and the hallways are the tunnels or "drifts." The drifts run from the shafts to the working areas, "stopes," where the ore is drilled, blasted, and removed. Haight explains that at Franklin, where the ore is thick, it is mined by horizontally transverse, or diagonal, stopes whereas at Sterling Hill, where the ore is in thin layers, or "stringers," it is mined by longitudinal or lengthwise stopes.

John Baum, resident geologist at Franklin Mine for thirty-two years, uses the analogy of a bookcase to explain the mining system in Franklin.

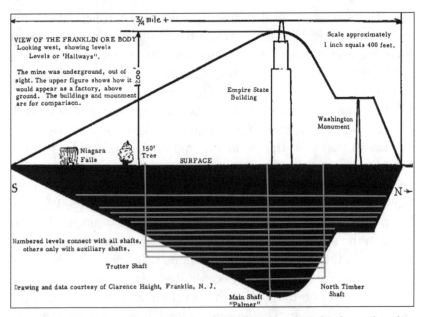

Figure 13. Franklin mine superintendent Clarence Haight drew this diagram of the Franklin Mine to be included in the *Franklin/Sterling Mineral Exhibit Catalog* of 1969. (Image: From the author's collection)

The mine, of course, is divided into sections, crosswise. In a way, it's sort of like books in a bookcase. Each book is going to be a working place. You mine out alternate books, fill up the space and then take the ores that are left. They are then called pillars. And those pillars align from the top down. That way, you get all the ore in the mine, every bit of it. There's no waste. Very efficient that way.

It's expensive because as you're working the final pillars, you have working places on either side that have been filled and you've got to keep that fill from running in and diluting the ore you're taking out. It means a lot of building of walls, that's exactly it. Driving flat slabs down at the edge to keep it up. Our mining method was a very safe one in that we were working from the top down. Every time a working place was finished, the miners would put down a plank floor so that when they came in underneath for the next ten feet, they had a wooden roof over head, so that was good. Every time they took out another ten or fifteen feet of ore, they had to put up another set of timbers.

Franklin miner Nick Zipco corroborates John Baum's claim that working from the top down is a safe method of mining. Here Nick compares the stope or work area to a living room.

Franklin, where I worked, was just like working in this living room. After I took one or two slices out, just like in this living room, I worked myself back. Take the sides off. Then I'd put timber, big timber we called sills, eighteen foot long and they're about a foot around. They're big logs. We'd put it by our set — this is what they called a set, the square room — as you're working, taking the ore out. Then, when you're ready to fire this place down when all the ore is taken out, then you start from the back end putting these big sills down. Three sills in between your set. Your set is about eight foot in between, a leg on each corner like in this room. Then what we do is drill each leg that is standing vertical. We drill a hole in there to put in a stick of dynamite. I can't fire it now until quitting time when all the men is out of the place, out of the pillars. When all are out of the pillars, then I fire the place down. Cuts them legs off and everything comes down.

I didn't mind Franklin, but I didn't want to work in Ogdensburg. I didn't like their operations, the way they operated. My brother Mike told me all those different things what they do down there. My brother Mike lost two fingers down there. He was drilling a block hole and a chunk come from the roof and cut two fingers right off. I did not like the operations in their mine.

If I went there, I would be working under loose all the time. Big boulders and stuff over my head, which I couldn't control. He puts pins in there. Big long pins. Then he wedges them tight. It's got to hold that up. If that's loose, maybe something else up above is loose. I don't trust it. I can't see through that rock. Franklin was a lot safer. When I worked in a place in Franklin, I put a big timber in, a prop we called it. When I started working in the working place pulling the ore out, I was working with timber overhead. Always had timber overhead.

Veteran geologist Robert Metsger explains the time and work involved in establishing a mining operation. He also contrasts the operations and the makeup of the ore bodies at Franklin and Sterling Hill.

Figure 14. Franklin miners navigating around stopes. (Photo: *Zinc Magazine*; Courtesy of Ogdensburg Historical Society)

When ore is discovered, the average length of time that it takes to hoist the first ton of ore out of the mine is seventeen years. That is true even today. In the first place, any ore body that's found now is not going to be at the surface. If they drill a hole and find ore, they have to find out what its extent is, so they have to drill a lot more holes, which takes a matter of years. You can't just use one hole. You might have the only piece of ore there is, so you have to drill it out and find its limits. Then, after finding the limits, you have to sink a shaft and you have to drive drifts or tunnels out along the ore until you've got an outline. Then once you've found out what the distribution of the ore is, you have to develop a mining system for it. You have to plan it so the darn thing doesn't collapse after you get in. Then, if you're out where there isn't much housing as there wasn't out here back then, then you have to build a town. You have to build a hospital. All of that takes time.

The Franklin Mine was such an old mine. When it started, back in the 1850s, you didn't have mechanized equipment like they do now. Ore had to be moved by hand from where it was blasted down to a vertical shaft for transport by car. That couldn't be done over a great distance so the Franklin Mine was honeycombed with raises or vertical shafts, that is, vertical openings in the mine. Franklin had been mined out so thoroughly that way that there was just this honeycomb of openings. Back in the early part of the twentieth century, why, it was considered that the mine was just about exhausted. There was still a lot of ore left, but they couldn't get it out so they hired Catlin, a mining engineer from South Africa. He came and first filled all the openings in the voids. Then he developed a system of mining where they could go in and get the remaining ore. By the time the mine closed in 1954, I'd say there probably wasn't more than a couple of tons of ore left at all. It was just about as clean a mine as it was possible to have.

At Sterling Hill, there was a larger ore body left. From a geologic standpoint, it was more interesting than Franklin because you could see more. There were quite a few problems at Sterling Hill that were not mining problems. They were geologic problems. They mined about twenty-four million tons from Franklin over its lifetime because all of the ore in Franklin was in one big block. At Ogdensburg [the Sterling Hill Mine], which was about half the size, it was much more difficult because the ore at Ogdensburg was in narrow veins and varied considerably. It was much more of a mining problem than Franklin.

Tom Sliker, who, in his thirty-seven years at Sterling Hill, rose from mucker or miner's helper to mine superintendent, felt that safety in the mine depended on the miner himself. In this section of his interview, Tom discusses "loose" — rock that may not be secure, "scale" — to pry free or separate rock using a

Figure 15. These miners at the Calumet & Hecla mine in Red Jacket, Michigan, are shown working in the mine that was described as having the deepest vertical shaft in the world. (Photo: Courtesy of Val Roy Berryman)

lever, and "roof bolts" — long metal rods used to secure the roof. Tom discusses the safety of roof bolts in comparison to scaling.

> I suppose ten years or so after I was in the mine, they came out with what they called a roof bolt. Before that, we used to have to scale all our loose. You sound loose by using the scale bar — metal hitting on the ground and the tone of it tells you where it was loose or solid. A solid tone would be solid ground. A hollow sound would tell you it is loose. The different sounds tell you how thick the loose is. You can have a piece of ore that's four foot thick; you get a certain sound from the bar, in the ring of the bar. If it's real thick, you get a lower tone from that bar. You done it by sound and scale it down. Use a hammer and wedge to wedge it down.
>
> Then after the bolts come, you were protected by putting these bolts in. The roof bolts were from three foot, six foot, nine foot, and twelve foot. You drilled holes in the ground and anchored these bolts. Each bolt would hold about five or ten ton. From there on we used roof bolts. You never done any walking under any open ground unless you put bolts in as you moved forward.
>
> Most of my work was done by development. I sort of liked development because every day you drilled and blast. You had virgin ground. Nobody ever seen it before but you. It's brand new. I went from mucker right to the top. When I quit at the age of seventy, I had run the mine for twenty years.

John Pavia, a Sterling Hill miner, started working there before roof bolts were in use. During the fifty years John worked as a miner, he saw many technological improvements, including the roof bolts. As John remembers:

> In Franklin, they work different ways. Pillars. Here in Sterling Hill it's stopes. The stopes are big. The pillar isn't big. We used to work from one level to the other one. In a lot of places, we didn't have no roof bolts. About four or five years before I left, they start to use them. They don't cost that much. The timber cost a lot of money.
>
> Here's the type of work I used to do in the mine, the drilling machine. The cars, the skips, I did all those jobs. I was in charge of the dynamite. I worked as a motor man. When they make the raise, they drill maybe four by four,

something like that, four by five. They blasted. Only one man. Years ago, it used to be two men. Then they bring in a different kind of machine, then only one man. The motor man, you fill up the cars. You had a helper. You had to have a helper in order to lift the planks out of the chutes where the ore's coming; you've got to have two men. Then you run the motor and dump.

Oh, what a change in the last ten years. Before we had the crank machine. Then we make ten, twelve holes a day. When I retired, they have the starter. Three machines, one man. Sometimes they drill maybe nine hundred feet a day. One man!

I was a timber man too. Always underground. I know I could go on the top, but I told them, I said, "Look, I want to retire in the mine." Because, you know, I'm used to that. On the top, it's different. Different kinds of jobs. I'm not used to that. I work in the mine, and that's it.

Clarence Case is another Sterling Hill miner who loved the work and "enjoyed every bit of the time that [he] worked there." Clarence tells of blasting and the danger of blasting. He felt that ceiling bolts were preferable to timber and perfectly safe.

The bolts, if they're put in properly, they're going to be all right. They're going to be very well. The timber sometimes can get in your way traveling, where the bolts are not in your

Figure 16. Miners moving broken ore with scrapers. (Photo: *Zinc Magazine;* Courtesy of Ogdensburg Historical Society)

Figure 17. Narrator Tom Sliker [right] with a fellow miner shown loading powder into holes. Roof bolts are visible overhead. (Photo: Courtesy of Tom Sliker)

way. If the bolts are put in properly, you're going to be relatively safe. You had to be careful. When you did your blasting, you'd come in the next day; you had to scale that ceiling. You had to watch yourself every time you set your machine up to make certain that the ceiling was clear. That there was no loose pieces up in the ceiling. The first thing we do when we go in the mine, we set up our machine. If you're drilling a stope — a stope is a room — it's a space about thirty feet wide. It could be ten, fifteen feet in height, and you're probably drilling the ceiling down. As that drops down, that gives you more height as you drill the next part of the ceiling. Once you've drilled these holes — you drill them usually about four feet apart — maybe you'll drill six, seven, eight holes, and you load those with dynamite, and you blast the ceiling down.

The next day, you'll come in and you'll just continue blasting the ceiling down. Eventually, they move the whole stope out. Then they fill it in. The process is filling it in with limestone, the slag that they've retrieved from Pennsylvania,

Palmerton. They bring back the slag in the railroad cars that took out the good ore. We have to fill it in to have room to work for the next time because if we didn't, we'd have all holes in the mine. It would look like a big Swiss cheese. I kinda liked the raise work — drilling fill raises. The fill raise is about four feet square and it goes up approximately at maybe a fifty-two degree angle. That's probably one of the most dangerous jobs in the mine because as you climb, everyday you're drilling nine feet. You blast out nine feet more until you reach the next level, which is 110 feet high. The dangerous part of it is when you're going up on the chain and the day before's blasting, a piece of loose could be hanging on the chain or the foot wall. As you climb, if you loosen that chunk, it's going to come down at you. You have to be extremely careful that you don't get hit with a piece of ore coming down, a loose piece.

The next day we would bring the chain up and tie it around the lagging. Lagging is about a four or five inch timber that you cut in steps across the hole so that you have something to step on. Then we'd do our drilling. The next day, we'd bring the chain up that much further. You look up the face of the raise; you have to climb that chain to get up to where we were before. You climb until your feet are hitting the foot wall. It's about a fifty-two degree angle incline. You're stepping on it; you're climbing it. Every day we put pins in the foot wall as you go up the raise. Each one of our drilling days was practically nine feet each day. Eventually, we would reach the next level by doing just that.

Bernie Kozykowski, who worked as a miner in Sterling Hill in the 1970s, provides additional comments on roof bolts and gives a clear explanation of the differences between mining methods in Sterling Hill and Franklin:

> Roof bolts were an advent of changes in the technology over time. It was coincidental as much as anything else. The timbering that was used in Franklin was used in pursuit of a recovery process developed by a mining engineer by the name of Catlin, who was hired by New Jersey Zinc. One of the most striking differences in the mining in Franklin and Ogdensburg — we used to call Franklin, Franklin, and Ogdensburg was called Sterling — when the ore was first

removed, it was rather haphazard in the first, oh I would say, twenty years of underground mining operations.

Then, from that point, for the next twenty years perhaps, what the mining company did was remove the ore body by taking a slice of the ore and putting a stope up. They simply mined from the bottom up and let gravity do its work. The ore dropped down and they would draw the ore out. This would be done in segments running from one side of the vein of ore, which is like a string through the ground. They would go from one side of the vein to the other; perpendicular to the direction the vein was taking in its natural state. When they mined out these stopes, they left pillars in between. If you go to the museum in Franklin, they have some very good examples of a stope and pillar mine. After the stopes were taken, then they went into mining the pillars.

That's where Catlin became very, very important. In essence, what they were doing, they left half of the ore behind so that the ground wouldn't collapse. That's when the timbering became important. If you go to the museum, they will give you a good talk on what they called the top slicing method that was pursued in Franklin.

There's a big difference between Franklin where they pursued the ore directly across the vein as opposed to Ogdensburg where they followed the vein. Where you have stopes and pillars, you have one problem. In Ogdensburg, you have longitudinal stopes where you actually followed the vein, sometimes with distances up to three hundred feet. The difference is that, in Franklin, the vein was very thick, especially toward the lower five or six levels. In Ogdensburg, they also did stope and pillar mining, but they did it only when the conditions at Ogdensburg were conducive to that type of mining. It was only in a very limited sense where the vein became very thick that Ogdensburg was able to do that. But in large part, the vein was very thin in Ogdensburg in comparison to Franklin. None of the mining that was done here would be used today.

One of the things they introduced in Ogdensburg that was interesting was something called hydraulic fill. When you recover the ore, you leave a big void in the ground. If that void becomes too large, it becomes uncontrollable. Both in Franklin and Ogdensburg the company spent an awful lot of time filling. Taking waste rock back, the tailings from the mills at Franklin and Ogdensburg are back in the ground for that purpose. In the late 1960s or early 1970s, the plant in Ogdensburg had produced the hydraulic fill system. The old methods required a lot of labor to handle the material and it wasn't particularly stable. And it was cumbersome.

What they started doing in Ogdensburg — I used to work in this section — they would actually use water to transport very fine fill material in the form of sand and fine gravel. They would pipe it all over the mine, water and sand together. Typically, the mine operations in Ogdensburg, for the most part, were undercut, meaning you go from the bottom to the top. They would mine out twenty feet of ore. Then they would drill holes into the vein, blast that down, recover that, drag that out. There's a lot of technical jargon as to how that's done. Then they would come in with a hydraulic fill crew and they would fill up not twenty feet, but ten feet and that would serve as a working platform. Sand and gravel, primarily fine sand and the last couple of feet, was mixed with cement to stiffen it. It wasn't a good concrete, but it served its purpose.

Figure 18. Miners construct a fill dump at the twenty-two-fifty level. (Photo: *Zinc Magazine*; Courtesy of Ogdensburg Historical Society)

Harry Romaine went to work in the Sterling Hill Mine in 1937, when he was twenty-one years old. He worked for the Zinc Company for the next seventeen years. During that time he held several different jobs but he never felt fear of either the work or of being underground.

I was never afraid to work in the mine. I never had any fear of the mine. Ogdensburg was always considered a safe mine, to a certain extent. I'd be in places down here working and the hole would be a hundred feet long and you wouldn't know what was on top. It never scared me one bit. I started out in the timber gang. We used to shore up the entranceways where you traveled. Up the side and across the top. Then we'd put in long poles wherever there was loose in what they called stopes. Some of the poles would be thirty, thirty-five foot long and we had to prop up the loose. It would be a big hole where they had taken the ore out and the loose had to be shored up because the men would be working down at the bottom to finish cleaning out. So they had to be protected.

Then we used to have to build the chutes where the ore would come out of these places into the tram cars. They was all made out of wood. There was a slide went from the five hundred level to the seven hundred level. That supplied all the timber and everything that the working people needed on these two levels. The slide, that's like a chute that you put the materials on. It had a wooden bucket on it and it had a motor and a cable and we used to send everything down on that slide.

The five hundred level was the main level for what they called the powder magazine. That powder magazine would hold anywhere from four hundred to five hundred boxes of dynamite, in storage. That's a lot. We used to average two truckloads from Hercules [powder plant]. They'd come up and they'd unload on top. Put it on what they called the cage. Send it down to that level, the five hundred level. That would be unloaded there and taken back in the magazine. Now part of my job was on the five hundred level while I was a pump man. The boss would come up about eleven o'clock from the different working places to the five hundred and he'd bring up a slip of paper to tell me how much powder they needed in each working place. Some places only needed five sticks. Some places needed two boxes. And I used to go into the magazine and get that all ready. Take it out to the main level of five hundred. Then after the bosses all through the mine went up for their lunch, then the man that operated that skip, he used to come down after taking them up. He'd pick up that powder and take it to the different levels like the five hundred, the eight hundred, the eleven. Then somebody would send it down to the other two levels on that slide. Every day we had to do that.

There was no excess powder scattered through the mine. It all had to be ordered. Just what they needed. When he'd go through, he'd say to the men, "How many holes you going to have today?" Well, they knew how much powder they'd have to have, and that's the way the order went. Just enough to do their job. But everything was kept on the five hundred level and all the fuses was kept on top, up on the main ground. They'd make them up there and put the blasting caps on. Then they'd deliver those to the working places. They weren't kept in the same place. That was a safety feature. No contact. And then before they started using the powder — maybe they didn't have all their work done before the powder come to their working place — so then, what they would do, the fuses was taken one place and put into a wooden box and there had to be at least fifty foot difference between the powder stored at the working place and the fuse. They didn't get together.

Miner John Kolic started working at Sterling Hill in the 1970s and continued working there until the mine closed in 1986. John speaks about the technology of the explosives and explains why he very much enjoyed working in the mine.

It was interesting. I liked working with the rock. Drilling. Blasting. Overcoming problems. In many cases, the job of mining — like in many other occupations — you have to overcome problems that you run into from day to day. At that time, 1972, they put you right in as a miner's helper and you relied on the miner you were working with. You learned from him. It was on-the-job training. A miner's job would be to drill holes for the purposes of blasting down the ore, okay? Now, you have to know how much space you can leave between the holes, how much powder you have to put in there. Blast the rock down and then you've got to get on top of that pile of rock and check the top for any ground that you

Figure 19. Miners remove explosives from the main powder magazine on the five hundred level. (Photo: *Zinc Magazine;* Courtesy of Ogdensburg Historical Society)

may have loosened up that didn't fall. You have to scale that down, and, if required, you put in rock bolts or timber supports or whatever else to make it safe. Then that broken ore has to be removed, dumped down a hole to a place where it can be loaded into a car that's pulled by an electric locomotive and hauled to the shaft so it can be hoisted up to the surface to the mill.

The biggest technological change that affected us as miners during the time I was there probably would be the explosives. DuPont stopped making dynamite. Relatively speaking, it's a hazardous substance to use. They developed a variety of what they called water gel explosives which were much safer. Much safer. Very difficult to set them off accidentally. So that was a good change. It was an interesting job. Getting the minerals.

Steve Misiur is in the unusual position of having worked as a miner at Sterling Hill after the mine actually had been closed for some time. Steve learned mining techniques from John Kolic, who was employed by the Sterling Hill Mining Museum to dig out a new tunnel for the mine exhibit. As Steve explains:

> How I got to be a miner itself is a rather odd story. I never intended to be a miner. I was here [at the Sterling Hill Mining Museum] as a volunteer, doing whatever work they wanted me to do here. Then Dick [Hauck, co-owner of the mining complex] gave me a job to remove the eleven hundred foot crusher. There were two crushers that they used underground. The first crusher — that's on the nineteen twenty level in this mine — was already flooded out because, in 1989, when Dick and Bob bought the property, the water had already risen up to the fourteen hundred foot level.
>
> So we tried to salvage anything from thirteen hundred on up to the surface that was underground. They got PR

Figure 20. A miner loads broken rock at Sterling Hill Mine. (Photo: *Zinc Magazine;* Courtesy of Ogdensburg Historical Society)

Figure 21. The ore is loaded into a car and pulled by an electric locomotive to the shaft to be hoisted to the surface. (Photo: *Zinc Magazine;* Courtesy of Ogdensburg Historical Society)

Engineering Company of Toronto, Canada, to come down here. They make and rebuild the rock crushers. So, as a source of income, they told this engineering firm that makes these rock crushers, "You can have the eleven hundred foot rock crusher." Fine, but the Canadians needed help to get this rock crusher out. So Dick says, "Why don't you work with these Canadians?" So I was hired by the Canadians to go down to the eleven hundred foot level, and basically disassemble a thirty-five ton rock crusher. I was paid directly by the Canadians. After the crusher was done, Dick evidently had enough confidence in me to want me to do more work.

One of the jobs that Dick gave me after the crusher was to work with John Kolic to blast out a new tunnel. One of the former miners that worked here that blasted out the Rainbow Tunnel with John had since moved on to other things. So John was left on his own without any help to do this new tunnel. I remember being in the mine and Dick was talking to John, and he says, "John, are you ready to start on the new tunnel?" He said, "Yeah, sure. But I've got to have a helper." And Dick goes, "Well, help is standing next to you." John looked around and looked at me. He shrugged his shoulders, "Well, okay." "So," he says, "We'll start Monday." Now remember, I had no practical mining experience before this. I was going to be working with John who had fourteen years'

Figure 22. Pump station on the ten-fifty level. The depth of Sterling Hill made it necessary to pump water from the mine. (Photo: *Zinc Magazine;* Courtesy of Ogdensburg Historical Society)

work. So we went down the tunnel, and he says, "Alright, you will watch and you will observe." Those were the first words out of his mouth when we started drilling. So he showed me how to begin running the drill. "And," he says, "You will change the steels." And I changed the steels that come out of the drill, and put them in.

Evidently, after half a day of working with John, I was ready to begin to operate the drill myself. How to lift it. How to hold it. How to turn the controls on. At the end of the day, John said to Dick, "Yeah, he'll be a good helper." The tunnel that we worked on is called the Dynamite Room. That was the winter of 1990. We did that in thirty-five days. Went 119 feet. And that was my first experience of mining. But, it was practical, on-the-job training, that 119 feet. The biggest problem that I can tell you about being a miner is it's a very physically wearing job. But I like being a miner in spite of the physical work.

All of the narrators who had worked as miners expressed a sense of achievement in doing such dangerous work. As former New Jersey Zinc Company employee and current Ogdensburg historian, Wasco Hadowanetz says:

> I think they were proud to work for the New Jersey Zinc Company. They were proud in what they were accomplishing.

1. Norris McWhirter, ed., *Guinness Book of World Records*, 282.

2. Clarence Haight, "Mining at Franklin and Sterling Hill — Synapse," *The Picking Table*, June 1960, 3.

When the Whistle Blew

At the peak of its activity, New Jersey Zinc gave employment to over 2,000 individuals. These were the people who drilled in the mines, worked in the plants and the offices and in all other branches of the organization.

Franklin Borough Golden Jubilee

In addition to the miners, the Zinc Company required several hundred additional and diverse employees to support its mining operations. From mill workers to the general mine superintendent, store clerks to office clerks, carpenters to electrical engineers, or nurses to the chief surgeon in the company-owned hospital, the number and variety of people hired by the Zinc Company to sustain the complex operation was remarkable. In the narratives that follow, we are presented with perspectives from our narrators in such diverse but essential jobs as those in health care, miners' support jobs, shop work, property maintenance, machinists, office work, and other jobs "on top." Whatever their job, however, the lives of employees were regulated to a great extent by the company whistle.

Ogdensburg historian Wasco Hadowanetz remembers the importance of that whistle:

> I remember there was a six o'clock whistle that blew. Oh, it was a shrill whistle. That thing must have wakened the whole town. It was to wake the miners. Then I would watch a parade of miners with their lunch pails going down Avenue "A," down Plant Street to the mine. We'd be walking to school and you'd see all of this busyness going on around the mill.

It was a large mill, about seven stories high, with all of this dust spewing out of it. The conveyor belts would be creaking and you'd hear the shake of the tables separating the ore, and ore cars creaking in the yard, moving out after they were filled underneath the old tanks. All of this hustle and bustle you'd be seeing on the way to school. Then the only thing you would hear on the way back, walking back from school, was the pinging underground where they were blasting. Probably from some of the levels that were nearer the surface. This limestone, this calcite, what they called Franklin marble, was so solid in most places they didn't need any supports. So sound traveled through the earth.

I went to grammar school here in Ogdensburg. Then I went to Franklin High School and graduated in 1948. I went to Newark College of Engineering a year and a half. Then, summers, Mr. Griggs, who was the head of the electrician's department, Franklin and Ogdensburg, offered me a summer job for me to work as an electrician right here in Ogdensburg. I worked for two summers in the electrician's gang. I helped put the signals in the new shaft in the mine, which was opened somewhere in the 1950s. I went all the way down to the bottom, which was 2,650 feet.

When you go in the adit here on the mine tour and you go to the end of that adit, that's what they call the top landing where the men went down in the mine. Next to it, or to the right of it, is a signaling system. They used to have what they called the "cage man," a man who was like an elevator operator. He would wait until the men got in the cage and then he had a key that he could turn and signal where he was and where he wanted to go according to the signal system. That signal was transferred to the "hoist man," up on top of the hill. He would lower them down to the level that they wanted to go. To the left is a compartment where all the wires go down and all the water pipes and all those things that are necessary to the mine. There's a ladder there that goes all the way down to the bottom. In case the hoist doesn't work, you could go back to the ladder. Sometimes the men, rather than wait for the cage, they would walk up the hundred feet to the next level they had to go. That was an alternate way of moving around, up and down between levels. But what I did was on those steps. You had to stand and put these wires down through and screw the conduit pipes on all the way down to protect the wires, 2,650 feet, all

Figure 23. The first cage would go down at about seven o'clock. (Photo: From the author's collection)

the way down to the bottom. Then you had to have outlets for each level where the signaling system is and the lights and so forth. The signaling system was for the shaft, that was one of the main purposes. The other purpose was to signal for the ore cars and to regulate the ore cars to go up and down.

During the time that I worked as an electrician's helper, one of the interesting jobs that I had was the weekly mine and mill electrical inspection. Our boss, Tom Dolan, sought volunteers to work Friday evenings to inspect the whole mine, the electrical equipment. Going down in the mine, we sat on the back of these ore skips. Our feet were only about four inches from the rails. The hoist man would kick in the clutch and we would go as fast as possible down to the bottom. One time the hoist man forgot we were on the skip and he hoisted us underneath the chute. Fortunately, the men that were there always looked down — they never just opened up the chute — and they noticed we were there. So they had to signal the hoist operator and say, "Bring it up to the level." We got out and then they filled it.

One other time the chief electrician and I were riding up and the skip went above the top level. In other words, the hoist operator thought he had brought up ore to dump out into the crusher. So, as we started going up above the top level, the chief electrician shouted to hold on to the cable. The hoist man must have realized that we were on the skip because he returned us to the top level.

In the following, Clarence Case relates his own memories of the company whistle.

When I was growing up, most everybody in Ogdensburg worked for the Zinc Company. You could see a whole bunch of men coming up with a lunch pail under their arms, automobiles going up the road. They're all from the New Jersey Zinc Company. Three-thirty when the whistle blew, these guys were all coming out in droves. Seven o'clock whistle blew, that was time everybody should be in their work. If you walked in — just in the gate — by the time that whistle blew, you probably were a few minutes late. Changing your clothes, that was your own time. You went down in the mine at seven o'clock. You had to be there early to get yourself changed. That's when your time started. Seven to three-thirty, that was company time. You could just about do your meals and your appointments by that whistle. They also had the fire whistle of the company down there. If there was a fire, they blew it a certain way so you knew there was a fire.

When I walked in the gate, I would throw my time tag in the window. I had to go through a turnstile. When I came out at night, I'd pick my tag up off the nail. Now, if that tag was not on that nail, that means they want to see me in the office. No reprimand. Nothing like that. It was just signing papers or to tell me what my job is or whatever. The tag was just a number. My number was twenty-eight forty-six. I was born in 1928 and I started in the mine in 1946. I haven't got anything to show that I worked there except my tag.

Until the organization of the Zinc Company, health care in Sussex County was limited. Only a few doctors were in the area, and the nearest hospital was thirty miles away. Following the end of World War II, Genevieve Smith returned to Franklin where she had grown up and worked at the hospital.

After service, I came here and worked here for eighteen years. I was working [in the hospital] for the New Jersey Zinc Company. After I worked there about a year, I went into the operating room and I stayed in the operating room as a supervisor for the rest of the time that I worked there.

Franklin Hospital was the first hospital in the county. Then Newton Hospital was started. Then Sussex Hospital. It [Franklin Hospital] was well equipped for the work that we did. When I was working in the operating room and

Figure 24. The new Sterling shaft was completed in 1951. An article in the April 1951 issue of *Zinc Magazine* proclaimed: "In addition to the regular bell signals and telephone system, a modern electronic signal system will make it possible for the cage man, who rides with the man-cage, to hold a two-way conversation with the hoist engineer at all times, even when the man-cage is in motion." (Photo: *Zinc Magazine;* Courtesy of Ogdensburg Historical Society)

> people would wind up on the operating room table, they were asleep, [but later] their head would go up and they'd say, "Are you here, Genevieve? Where are you?" They liked having somebody there that they knew.

In 1922, Alvah Davis began working in the Franklin Mine as a machinist helper, a job that paid only twenty-six cents an hour. When he was promoted to hoist engineer, he was paid a considerable amount more. Here Alvah describes the responsibilities of a hoist engineer that warrant an increase in pay.

> After probably twelve years or so, I was trained to be a hoisting engineer. That involved running a steam engine. You used to lower the men in the mine. There was two engines. One was just for ore alone. And one for just handling

men and supplies, and timber and so forth. That was the one
I was on. I trained for both engines as far as that's concerned.
They were both the same, but I didn't run much on the ore
hoist. I was more on the men. You had thirty men's lives in
your hands when that cage is going down mornings and
coming up. It was supposed to be one of the best paying jobs,
as a worker. There were engineers that got the same amount,
but for a working man, it was the highest, I guess. The
engineers in the power house made the same rate. It was a
little bit higher than the top machinist for instance. But it
was a responsibility. No question about it.

Certainly not everyone could stand the strain or responsibility of being a hoist man. In fact Steve Sanford, who worked as
a miner at Sterling Hill in Ogdensburg, refused the job.

> They asked me to be a hoist engineer, but I couldn't stand
> it. Up and down all night long. You had to be so careful to do

Figure 25. At quitting time, "You could see a whole bunch of men coming
up with a lunch pail under their arms." (Photo: *Zinc Magazine;* Courtesy
of Ogdensburg Historical Society)

it just right. Sometimes hoist engineers would get to daydreaming and sometimes the cage would lock. They had automatic fail-safe on it. If the fail-safe detected a slack in the cable, it would clamp down these clamps on the beams of the shaft and the cage would freeze. Then the shop man would have to come and unlatch the cage and get it free again. I'm not sure what the base rate pay for the hoist engineer would have been, but I'm sure they made good money because people made a living at it and supported households with it.

Some employees preferred to work "on top," that is, aboveground, because work inside the mines was too dirty. In the following, Sterling Hill geologist and former miner, Bob Svecz comments on the clothes they wore and the difficulty in keeping clean.

If you worked underground, you really didn't take your clothes home to wash them. Within ten minutes you're filthy dirty. Basically, you'd wear your clothes until they fell off your back. You didn't care what you or your equipment looked like as long as it worked. Equipment didn't have any fresh coats of paint on it. It worked, that's all we cared about.

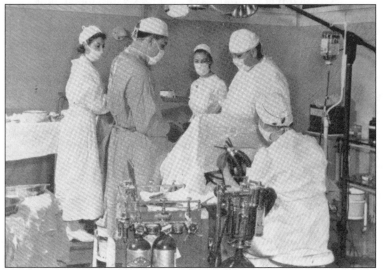

Figure 26. The operating room of the Franklin Hospital in 1954. (Photo: *Zinc Magazine;* Courtesy of Ogdensburg Historical Society)

Underground, it's almost like everything looked the same color. All rusty or muddy. Unless you had a mineral light, then everything would light up like a Christmas tree.

In places, we had to wear the rubber boots because of all the slop from the hydraulic fill and climbing up and down the ladders your hands would get dirty. I never wore gloves for one simple reason. If you had to take notes, you had to take your gloves off to write the notes. You'd walk away from where you put your gloves down taking notes. You don't have your gloves. You've got to run back and pick them up. You're always leaving them someplace, so I decided why bother with the gloves.

They didn't set any fashion show here. It didn't matter. It didn't matter if you took a shower, although your buddy would appreciate it if you did. One person I know of who did work here, his nickname was Brown Cloud. It came from a cartoon. He was a character that was always walking in a cloud of dust. We had a guy here like that. Never took a shower. It was okay if he kept downwind of you.

In his interview, Clarence Case also addresses the issue of keeping the uniform clean. He adds:

A lot of miners would leave their locker and their clothes hanging up. They wouldn't take them home to wash. A lot of them would just leave them down there till they wore out. My mom always told me, "If you don't bring your clothes home every two weeks, then bury them in the mine." So I brought my clothes home for laundering. You had to buy your work boots, your hat, your coveralls. The oil skins, that, they gave you. The company supplied the oil skins. That was a great big rain coat and rain pants, you might say. When you worked on the machine, working on a steep angle, the water from the machine would come down on you. You had to have the oil skins. Over the years they got misplaced, lost, and so on and so forth.

Another aspect of mining that made people prefer to work "on top" rather than in the mine was the danger involved in the work underground. Bill May's father strongly advised him against working in the mine.

Figure 27. An employee of the Franklin Mine at the controls of a steam-driven ore hoist. (Photo: *Zinc Magazine;* Courtesy of Ogdensburg Historical Society)

Figure 28. A miner operates the hoisting engine on the eighteen-fifty level at Sterling Hill Mine. (Photo: *Zinc Magazine;* Courtesy of Ogdensburg Historical Society)

My father didn't want me to go into the mine. He said, "Too dangerous." He worked in the mine when he was a young man. His best friend got killed in the mine. My mother said to my father at the time, "You get out of the mine. I don't care what you do, but get out of the mine." The danger in the mine was an average of two men a year killed.

I ended up getting a job in the Zinc Company store. November 1946. The hours were seven-thirty to five o'clock, I believe. We had half a day off on Wednesday. Then, Saturday, full day. I was making a hundred and ten dollars a month. That was the salary when I started. It went up. I think I ended up with a hundred and sixty dollars. Something like that. I stayed there until the store closed. I think that was in 1951.

I was a clerk in the store. Then as the older men retired — there was a fellow named Doc Holshizer in there — I more or less took over his position. The manager was Elsworth Wright. He passed away. Adam Gorsczyka came down from

Figure 29. "When you worked on the machine, working on a steep angle, the water from the machine would come down on you. You had to have the oilskins." (Photo: *Zinc Magazine;* Courtesy of Ogdensburg Historical Society)

the main office. He was an accountant there. He came in as the manager. He told me, "Bill, I don't know anything about the store business." He says, "I'll take care of the business end, you take care of running the store." So that's what made it nice working too. He, more or less, gave the people some responsibility.

They had a grocery department, a meat department — a butcher shop. As you came in the front door, to the right was dry goods. There was a tobacco counter right there. They had steps going up and down. Downstairs was hardware, paints, appliances, electrical appliances, toasters and things like that. Galoshes, that you don't see sold anymore. All kinds then, for men, women, and kids. Upstairs was the same thing, men's, women's, and kids' shoes. And clothes. Everything they handled was of a better grade than you could buy in most of the stores. We sold from soup to nuts. We had lawn furniture. We sold mattresses, springs, linoleum, window shades, venetian blinds.

What I remember about the store was you got to know everybody in town. For lunch, I used to go up the street. There

Figure 30. Hoist engineer at the controls of the main shaft hoist. (Photo: *Zinc Magazine;* Courtesy of Ogdensburg Historical Society)

was a little luncheonette. It was run by a guy they called "Ma" Wilson, an old timer around town. Then you would find Chief Irons in there. Herbert Irons, who was the chief of police in town for years. He would have lunch in there. Our boss, Elsworth Wright, used to stop in there.

Never heard of coffee breaks at that time. Elsworth Wright might go off in the morning for a coffee or whatever. But not us. We didn't have any coffee making in the store then. Nobody had that. Every place you go now, every office has got the coffee maker, but we didn't have that. If we had anything, it was stuff that was in the store. We'd help our self to a bakery product. We sold soda. We had ice cream. That was about it.

Everybody that worked in the store was salary. Whether they thought because I wore a necktie and a shirt, I don't know if that was where the line was drawn. [laughter] We had like a white-collar job. We weren't getting our hands dirty like the guys in the mine. Maybe that was the difference. But, even in the mill, you still had salaried men. The head of the different departments or whatever.

By the end of his career, John Naisby was one of those salaried employees to whom Bill alluded. However, when he first began working for the Zinc Company he did a number of jobs in the various shops in Franklin. Here John describes the various training he received before working his way up to department head at Sterling Hill.

[The company] sent me down to the machine shop to get experience. Then, after I was there for a year or two I guess, they were doing some renovating in the boiler house. They were putting in all new boilers. Three large boilers in place of the ten that they had. Well, that meant high pressure welding so the company sent me out to the Lincoln Welding School for a month so that I could come back and set up tests, for testing our welders in Franklin to qualify them for welding high pressure steam pipe. So that took me a while to accomplish. I had to design the equipment to make the tests on the weld and that sort of thing. That involved also welders from Ogdensburg. We had quite a few excellent welders as a matter of fact.

I was sent to work under Ben Percival. He was the foreman of the plant down in Ogdensburg. I worked under him, assisting in renovating the machinery and that sort of thing. Then later on, we started building a whole new plant down in Ogdensburg. I was involved in the supervision of the construction of the machine shop and the hoist house and mill buildings.

In Franklin, I worked by the whistle. Then, when we moved to Ogdensburg, I brought the clock down that operated the whistle and I set it up in the office there and got it all rigged up ready to use, and the guy that was the superintendent at the time, he says, "We're not going to have any whistles blowing." This was a case of this fellow being sent here to cut down, you know, and get rid of some of these things that weren't necessary. Sterling Huyett. They sent him here from Palmerton to straighten things up, but I was fortunate in having men that were skilled and men with a lot of experience. They were all skilled workmen. I don't know of any jobs that were not well done. Of course, that was my job to see that they were. When I left, I was chief of construction and maintenance, I think it was. That was the way that they got around paying you more money. Give you a title.

The Zinc Company's real estate division was responsible for keeping the company's buildings in good repair. For this purpose a crew of carpenters, painters, plumbers and laborers was kept on the payroll. In addition to maintaining buildings, the real estate division was responsible for selling property and collecting rents. After graduating from Franklin High School in 1930, Steve Novak was hired by the Zinc Company where he came to work in several departments.

[My first job was] with the real estate department, working on the outside. They took care of the grass and keeping the property shipshape and nice. My first job was as a janitor. Then I got a job and they transferred me to the time office as a time clerk. I didn't care too much for working inside an office so I got transferred into the millwright gang. The millwright gang was a group that repaired machinery and did building and construction. Just about all the fellows in the millwright gang were laborers, but they were all very good. They had to read blueprints and had to do construction. For fellows that only had a high school education, they did a remarkable job. They practically built all those buildings for

Figure 31. Sterling Hill employees make repairs in the machine shop. (Photo: *Zinc Magazine;* Courtesy of Ogdensburg Historical Society)

the Zinc Company. The mill and all those buildings. They had to do all the repair work, construction work, iron work.

One of the oldest of the narrators, Bill Dolan, started working for the Zinc Company before the company had running water. When he was barely sixteen, Bill started working at Sterling Hill, carrying drinking water to the miners. After several years, the Zinc Company gave him the opportunity to gain more education. The company paid for Bill to attend night school in Franklin, where he learned to be a machinist. Here Bill describes the early years with the Zinc Company.

> [I worked] around the shop doing odd jobs, small jobs. I had all good bosses. Of course my father was pretty strict with me. I worked in the boiler house, air compressor, hoisting house, in the mine on the pumps, and in the machine shop. The last few years, maybe the last four years, I was in the machine shop all the time. I didn't have to go in the mine. The others

went. We went to work at seven until half past three. You had to take the work as it came. Dirty job or a clean job, hard job or a easy job. You had to take it as it came to you.

Seventy-five years ago, August the fourth, I started to work for the Zinc Company. Those first couple of winters that I went to work for the Zinc Company, you would walk in the snow up to your knees. No snow plows in those days. Boy, that was cold. They were good to me. I worked steady. I lost very little time.

Steve Revay is yet another Zinc Company employee who started working in Franklin as a laborer while still in his teens. Steve relates:

I was only fifteen years old when I started to work at thirty-one cents an hour. The Zinc Company was a good place to work, but they never paid no wages. Their wage scale was very low. I did mostly pick and shovel. I worked out in the yard. On top. In the mill. I didn't work in the mine. They paid better in the mine, but they had an opening in the yard so they put me in the yard.

Figure 32. A crew from the real estate division paints one of the company houses. (Photo: *Zinc Magazine;* Courtesy of Ogdensburg Historical Society)

They probably wouldn't have allowed me in the mine at fifteen. I got to working in the sampling department. You took samples of ore all through the mill. As they processed the ore, we'd take samples here and there and all over the place as the ore went through the mill. We took them to the sample house and then they went to the laboratory to be tested to see how they were. They'd take the iron ore out, the willamite. Separate it. I'd take a sample before they separated it and afterwards to see how the separation was. Whether it was good or bad. All that kind of stuff. Sampling. I worked there about half the time.

The other half I worked out in the yard. Loading the ore in the cars. You had the railroad cars. They had the tracks and you bring the cars down under the bins. Then you'd open the hopper and load the cars for shipment to Palmerton. But there was no way to get ahead. That was the big detriment. That's the only fault I've got about the Zinc Company. Well, there's a lot of faults, but the one big fault was they were very, very low paying people. And it wasn't that they didn't have money because they were filthy with money. The Zinc Company was very frugal.

One complaint that was voiced over and over again by the narrators was that the Zinc Company did not pay good wages.

Figure 33. Sterling Hill employees load rail cars with the help of a locomotive crane. (Photo: *Zinc Magazine;* Courtesy of Ogdensburg Historical Society)

Figure 34. Coal miner opens a powder keg. Though working in a zinc mine was dirty, it was not nearly as bad as was work in the coal mines. (Photo: Courtesy of the Bureau of Mines Collection, National Mine Health & Safety Academy)

Figure 35. None of the narrators began work as young as eight years old, which often was the case in many of the Pennsylvania coal mines. In the early years of the Twentieth century, children were made to pick the slate out of the coal with their bare hands, which could result in loss of fingers. (Photo: Courtesy of the Bureau of Mines Collection, National Mine Health & Safety Academy)

However, in spite of this criticism, the majority of those inter-viewed feel that the company did offer many benefits that helped to compensate for the low pay. But even if people were unhappy with the Zinc Company, very few, during the difficult years of the early part of the century, would have had the temerity to quit. Only the very young or very confident would have thought of giving up a secure job. However, this is exactly what Bob Shelton, who was both young and bold, did. Ninety-four-year-old Bob Shelton is best known in Franklin as a director of the Sussex County Trust Company, the local bank. But, as Bob ex-plains, his very first job was for the New Jersey Zinc Company.

My father was a sick man and when I was fourteen, I left high school and got my working papers. I had to get working papers and I went to work in New York. And through some magic, I ended up with a job with the New Jersey Zinc Company as a call boy. Ran errands and carried things around. They were in the top floor of the old National City Bank Building, which was 55 Wall Street. Then they built their own building down in Maiden Lane and Front Street, and we all moved down there. They owned that whole building. Let's see, I went to work for them in 1916. It was about 1918 that they moved to their own building on Maiden Lane and Front Street. This was my first job. I was an office boy.

Then I worked in their sales department writing up vouchers for sales that they had made. I did that for about a year. Then I was moved to the credit department where I worked until about 1925.

When I left high school, I had taken night school courses in one of these business schools where I learned bookkeeping. I was an expert bookkeeper, believe me. I could look at a statement and could tell you whether it was right or wrong or what the trouble was. I had that kind of mind, you see.

But I couldn't make ends meet with the money they paid. A man by the name of Harry Heaviside, who later became the credit manager of the Zinc Company, said to me, "You know, there's a small bank up there in Franklin, and they don't have a credit manager. They're looking for somebody who has training in credits."

I came up here then in 1926. In October 1926. My God, that's a long time ago. So I worked in the bank. I worked for nine different presidents in this bank. The Sussex County Trust Company.

The Zinc Company was an old steady company, but I couldn't make ends meet with the money they paid.

In spite of the many complaints about low wages, most of the people who had worked for the company were glad to have

Figure 36. The Sussex County Trust Company on Main Street, Franklin, New Jersey, where narrator Bob Shelton worked for many years. (Photo: *Zinc Magazine;* Courtesy of Ogdensburg Historical Society)

had that experience. Whether they worked with drill or shovel, wrench or saw, pencil or thermometer, the majority recalled his or her service at Franklin or Ogdensburg with satisfaction. They seemed to share a sense of a job well done and would have agreed with Wasco Hadowanetz when he said that people took pride in their work and being employed by the New Jersey Zinc Company.

Hard Hats and Safety Boots

For many, the image of coal conjures up the grief-stricken faces of miners' families at the site of a mine explosion: Monongah, Castle Gate, Centralia, Farmington, Scotia — the list is long.

US President's Commission on Coal

Part of the public's continuing interest in mining is its fascination with danger. The thought that one could be buried alive or caught in a massive explosion a mile underground captures the imagination and sends shivers down the spine. Unfortunately, the history of mining is rife with ghastly accidents and scores of lost lives.

The worst mining disasters in the United States have been coal mining explosions, and among the most lethal of these were the explosions that occurred at:

Scofield, Utah with 200 deaths, May 1, 1900

Monongah, West Virginia with 362 deaths, December 6, 1907

Jacobs Creek, Pennsylvania with 239 deaths, December 19, 1907

Cherry, Illinois with 259 deaths, November 13, 1909

Centralia, Illinois with 111 deaths, March 25, 1947

West Frankfort, Illinois with 119 deaths, December 21, 1951.

The government must bear some responsibility for dangerous mines. It was only after hundreds of lives had been lost that the government began to enact safety laws. The Bureau of Mines was established only after three terrible disasters in 1907 — 84 men died in January at Stuart, West Virginia; 362 deaths occurred in December at Monongah, West Virginia; and another 239 miners died in the same month at Jacobs Creek, Pennsylvania. However, the Bureau was only a research and fact-finding agency. It could make recommendations but had no authority to impose safety controls. It would take an inexcusable forty-four additional years and many more deaths before the government would institute better safety controls.

After the tragic explosion at the Orient No. 2 mine in West Frankfort, Illinois, Congress enacted the Federal Coal Mine Safety Act of 1952. This act expanded the responsibilities of the Bureau of Mines to include the inspection of coal mining and the authority to close mines that did not conform to the safety standards.

Still, it took nearly another ten years before the safety regulations were applied to mines from which metal and non-metal ores were extracted. These mines posed a very different safety concern than did the coal mines; instead of explosions, the most imminent threat to miners working in hard-rock mines was falling rock. As a matter of fact, individual accidents from falling

Figure 37. Mining meant hard, dangerous, backbreaking labor for little pay and, because of the gasses associated with coal, coal mining was particularly dangerous. (Photo: Courtesy of Val Roy Berryman)

Figure 38. Hearses take caskets to the mine and coffins line the streets of Monongah, West Virginia, where 362 miners lost their lives. (Photo: Courtesy of the Bureau of Mines Collection, National Mine Health & Safety Academy)

rock caused more fatalities than did the explosions in the coal mines. Although the grim drama of large explosions was featured in the news and made the public aware of the dangers of mining, it was the frequent deaths due to rock falls and cave-ins that drove the fatality numbers so high.

The New Jersey Zinc Company made the well-being of its miners a top priority as early as 1912 when it implemented a vigorous mine safety program at Franklin and Ogdensburg. By the 1920s, the Zinc Company was using hard hats imported from England, and in 1929, it made the wearing of hard hats compulsory in all of its mines. The safety practices of the company in those early years were among the best on record. In 1969, the company was one of six mining companies to receive the *Sentinels of Safety Award*, a national recognition presented jointly by the National Mining Association and the Mine Safety and Health Administration.

To further encourage safety awareness, the Zinc Company established a division that was responsible for education, accident prevention, first aid training, fire prevention, and mine rescue activities. A segment of the mining population was constantly trained in safety measures and mine rescue, and to keep employees aware of the need for being alert and practicing safe mining procedures at all times, the company provided each employee with a copy of the book *Safety Rules and Regulations*. The safety division also published *Safety Age*, a monthly booklet that featured safety hints, first aid instructions, and analyses of all "Lost Time Accidents" of the previous month. In reviewing the ten accident cases for April 1939, the May issue of *Safety Age* gave the employee's name, tag number, cited the number of days lost due to injury, explained exactly what had happened, and determined who or what was responsible for the accident. In an accident which resulted in a bone fracture that would mean ten to twelve weeks of lost time, the evaluation states, "Nagy is a machine runner of long experience; however, he did not use good judgment in this case."

While making an example of an injured miner may have been considered a useful way to prevent future accidents, the Zinc Company was actually following a long established tradition of minimizing or deflecting its responsibility. If the miner was the cause of accidents, the company obviously was not culpable.

Robert Metsger, the last superintendent at Sterling Hill, confirms both the Zinc Company's commitment to safety and its superior safety performance.

> They were very safety conscious. They had a very good safety record. There were people that were killed, as in any heavy industry. But the safety record was far better in the mines than the safety record of people working on road cuts and things like that. Similar work but on the surface. I would say the Zinc Company had a very good safety record.
>
> We always had inspections. Not just the state, but the federal government. They would make surprise visits through the mine. We were inspected by the Bureau of Mines, and the Mine Safety and Health Administration. Also the water that we discharged into the river had to be monitored once a month and analyzed for heavy metals and for bacteria. That's an interesting story. We didn't pump very much water out of

the Sterling Mine. It was not very wet, which meant that our pumps on the eighteen-fifty level only ran for a couple of hours at night when power costs are cheaper. We would discharge water from the mine every night into the Wallkill River. If the inspector was coming around to inspect the water, since we only pumped at night, they had to call me and tell me they were coming so that we could turn the pumps on and they could get the sample.

Well, one time, there was a girl who was doing the sampling for the state. I guess she didn't know that she was supposed to call because normally you don't. It's *supposed* to be a surprise. But she didn't know that in this particular case you had to call. So she came and there was a little water trickling out of our discharge point by the river and she sampled it. The pumps were not on, but she didn't know that. She'd never been around before. When the sample was analyzed, it had something like a two thousand count of fecal bacteria. We heard about that right away. I couldn't understand how this was possible because all of our sanitary facilities underground consisted of portable toilets that were brought out of the mine. There was no way for that to happen. Finally found out that she had come and sampled without telling us and what she had sampled was water that we were pumping out of the Wallkill River up to the top of the hill to run through our compressors. Then it ran right back down to the Wallkill again. So we sampled the water in the Wallkill and it had the same count as what she had. Actually, we always had a zero count because the mine was back filled with sand and we used a lot of cement, which is limy so the water going from the mine, as far as bacteria was concerned, was very pure. It had no count at all. But, yes, we were inspected.

For example, one of the regulations in a mine is that you can't have anything loose hanging over your head. Everything has to be cleaned down. One of these mine inspectors went through with the superintendent who had been giving him [the inspector] a lot of flak. There was a ledge on the side a little higher than this table, about eighteen inches, and there was a rock sitting on the ledge. The mine inspector went over and he pushed the rock off and he said, "That was loose," and we had a violation. [laughter] But they had a very good safety record.

Several of the narrators mentioned the safety engineer, the

seminars on safety, and the safety regulations that were enforced. For instance, John Naisby, who was chief of construction and maintenance at Sterling Hill, declares:

> [The Zinc Company was] always excellent on safety. They had, of course, a safety engineer, who conducted classes in safety and first aid, and anything that had to do with safe work. It wasn't optional. I know I had to attend a few of them. Of course hard hats. Everybody had to wear those. They were pretty tough on the safety business. I know that the Zinc Company operations were considered to be a pretty safe mining operation.

Many narrators remembered the monthly publication *Safety Age* and the book *Safety Rules and Regulations* that was issued to all employees. John Baum, former resident geologist in the Franklin Mine notes:

> Everybody, the day they came to work, or even applied for a job, was issued a booklet of rules. It was necessary for them to sign the last page, tear it out, and hand it in when they went to work the first day. Rules like wearing hard hats and steel-toed shoes. They turned it in to the mine office, which is where the employees checked them in and out each day. A miner was given a check, which in this case was a round brass tag. It hung on a section of the wall. He went through this passageway and he took it from one section to the other. That showed he was there. Somebody from the time office would collect all those tags showing the men were underground. They knew just who was there and who wasn't there.
>
> The Zinc Company, of course, took care of its own. If the miner got hurt, the company put him in the hospital and took care of him. Of course if a miner got sick or had the flu or something like that, that was on him. He just lost time from work. There was unemployment insurance through the state. But if anybody got sick in those days, they got sick and paid the doctor themselves. This was just natural throughout the nation as far as I know. The Zinc Company built the hospital. That was the first hospital in Sussex County. If someone was injured, the doctor came. If he was injured in the mine, the doctor came down into the mine. The doctor was hired by the Zinc Company. These were good doctors. Surgeons. They

took care of me. I got injured on the job. Not very seriously, but that was taken care of. The hospital service was open with very reasonable fees, prices, to all the employees.

∾

Former Sterling Hill electrician Wasco Hadowanetz also remembers the Zinc Company's emphasis on safety, the monthly safety booklet, and the disciplinary action that would result from breaking the safety rules.

```
                                    Case Number

                                    Date of Report Aug.1, 1947


     INVESTIGATION OF ACCIDENT TO    Chester Ziiuch #270

               INVESTIGATED BY       H. J. Hegenbart, Safety Eng'r.

               DEPARTMENT            Franklin Mine; 940 Pillar, 70 Feet above
                                                            1060 Level
               SHIFT BOSS            John Rowe

               DATE AND HOUR
                 OF INJURY           July 28, 1947 - 5:30 P.M.

               WITNESSES             John Kotar #261

               NATURE OF INJURY      Contusion right foot



               TIME LOST

     HOW IT HAPPENED:-              Zdiuch, a helper, says they were
          scraping ore on the pillar slice and it became necessary to
          move the scraper block to a new position.  In removing the
          block to its new position he noticed some loose crushed ground
          on the face.  Zdiuch then put the block down and turned to get
          a bar to dress it down when the loose ground fell, striking his
          right foot.      He was taken to the hospital at once for examina-
          tion and treatment and was not released for work.

     COMMENT:                     The ore in this section is crushed
          considerably and must be constantly examined and dressed down.




     RECOMMENDATIONS:




     WFZ-CLH-JR-RJH-LH- T.W.B3                CHIEF OF SAFETY DIVISION
```

Figure 39. The Zinc Company kept very precise records of all accidents. (Image: Special Collections & University Archives, Rutgers University Library)

APRIL LOST-TIME ACCIDENT RECORD

10 CASES

NO. 460—CECIL LADUTKO
Laceration of Nose
Lost 4 Working Days

Accident occurred 1-5-39. Lost time in April due to repair of scar on nose. He was removing a piece of 4x4 timber from behind set timbers in pillar. When he cut one end free, the stick flew back and struck him in the face, knocking him to the ground.

SUGGESTION:—There was no wrong practice observed here except that man failed to consider that lagging might fly back when cut off.

NO. 318—HAROLD STANABACK
Lacer. and Fracture Tip 2nd lt. Finger
Lost 4 Working Days

Stanaback, a machine runner, was drilling in the raise with a stoper machine. When starting a hole, the machine slipped off drill board and caught his finger against side of raise.

SUGGESTION:—The only suggestion for avoiding this kind of accident is the use of more care and the watching of the machine all the time.

NO. 178—SAM SHULTZ
Fracture Tarsal Bone, Left Foot
Lost 10 Working Days

Shultz, a mucker, was barring down loose in the pillar slice when a piece fell and struck his left foot.

SUGGESTION:—Apparently Shultz stood too near the ground that he was cleaning down. The use of a longer bar would probably have prevented this injury, since the man could then have stood farther away from the loose ground, and in case a piece fell, he would have been in a better position to move away from the falling slab.

NO. 304—PIERRE McCARTHY
Fracture 12th Dorsal Vertebrae
Probably Off Work About 9 Months

McCarthy, Jos. Elchin and Herb. Thomas, were working in the pillar slice, starting to scrape ore to the pillar raise. McCarthy and Elchin stood near the hanging wall under a row of timber sets; Thomas was near the foot wall of the pillar, about 30 feet from the others, running the scraper hoist. Without warning, the timber sets collapsed and the fill and timbers buried McCarthy. Elchin had been narrowly missed by the cave-in, and as he and Thomas hurried to help Mc-Carthy, the fill above started moving again and there was another run of the chunks of rock inth the slice. McCarthy was released after about two hours work.

Figure 40. Even when there was "no wrong practice observed," blame for accidents was placed on the miner's failure "to consider" what might have happened. (Image: From the author's collection)

They were very sensitive to safety. They had a little booklet called *Safety Age*, in which they would relate accidents that happened. I know my father was very safety conscious. One of the things they didn't want you to do was carry matches. They didn't want you to smoke. If they caught you smoking, the first time, they would give you a warning. The next time, give you a day off. The third time, they'd fire you. So people were well aware of this.

Then they had safety courses they gave. There were groups of men that were on safety crews. They had teams and they tried to explain to the miners to be safety conscious. Actually, I remember one time my father had a broken foot. Even though they had steel toes on their shoes, a piece of loose rock came down and broke his foot. There were other accidents in the mine. What they would do in those cases, they'd give them a light job to do for the rest of their lives. Workers never sued at that time. Some people were killed in the mine. There was one man on the cage, stuck his head out too far and was killed. There was one man that was crushed by motors coming through a narrow tunnel.

Geologist Robert Svecz, who was working at Sterling Hill when the mine operations were shut down in 1986, recollects the rule against smoking in the mine, safety glasses, and his mining accident.

Figure 41. A fire-fighting crew wearing oxygen masks during a training session in the Franklin Mine. (Photo: *Zinc Magazine;* Courtesy of Ogdensburg Historical Society)

The biggest cause of not smoking underground — not so much here in Sterling, but in Franklin — they used a lot of timber. They were afraid that a guy smoking would just toss a cigarette. Well, they did have a fire in the Franklin Mine. It was a very stubborn fire, back in the twenties or thirties. It wasn't enough to shut the mine down, but they couldn't put it out right away. When you think about it, before they had the electric blasting caps, they used burning fuses. You had to light them. When we were mining here, we were drilling holes in the back, in the ceiling above your head. The holes are about a foot-and-a-half apart. We're using burning fuses to set the explosives off. You had carbide lamps at the time, something like a two-inch flame shooting out of it. It was very handy. You used that to light the fuses. Now you're lighting all these holes in the ceiling above your head. You have the fuses hanging down and dangling around the top of your head. And what's on top of your head? The lamp with the two-inch flame! And yet, you were not allowed to smoke.

If you didn't do your job right, you could be fired. Simple things. If you weren't wearing safety glasses, you'd get three days off for that. Or even five. You could always tell the people who didn't wear the safety glasses and the people who did wear them. If you're drilling, you're using the water to wash the cuttings out of the hole and to keep the dust down, and it's just spraying all over the place. It's just muddy water, so your clothes are going to get filthy dirty, and your face is going to get black. You could tell when they weren't wearing the glasses because when they'd get on the cage at the end of the shift their eyelids would be black. Whereas, if you were wearing your glasses and you took your glasses off, you'd have these light rings around your eyes. That would be clear. But most of the people watched the safety.

Smashed my finger once. Nothing that was my own fault. As a matter of fact, that probably saved my life. We were drawing all of the ore out of the ore chute for final clean up. We had gotten what we thought was the last car of ore out of the bottom of the chute and we had to clean out what muck was lying in the bottom because we were using hydraulic fill at that time. The lime from the cement would precipitate out on the bottom of these chutes and whatever was left there, it would cement in place. If you kept on doing that through each cut, it would build up and eventually block the chute off. So I had to climb up into the chute. I was going to get in the bottom to clean out the chute. I had to step on something so I had the guy close the gate so I could step on that, the roof of the gate, and climb in. To keep my balance on that car, I had my hand on the side of the chute, and I told the guy to close the gate. I actually stepped out of the way so the gate, when it closed, wouldn't hit me. But I left my hand on the side of the chute and it came up and smashed my finger. Same way as though I'd opened my car door and you stick your hand in the doorjamb and I slam it as hard as I possibly can. Well, of course, my finger split wide open. So go out to the station, get the cage, go to the hospital. They scrubbed it and sewed it up. The guy I was working with, when he walked back to the chute, the chute was full of muck. What had happened — there was a flat spot in that ore chute a way's up and the ore had hung up on the bottom side of that chute, the side of it, but it was open on the top. Now, I would have been standing in the bottom of that chute when it came down and there would be no way, no time for me to get out of the bottom. About four tons. It would have come down on me. So I look at it like this: "I smashed my finger, but in a way it did save my life."

Not all accidents had such fortunate results. The use of explosives was extremely hazardous and often led to casualties. While they may not have been directly involved with accidents associated with explosions, all who worked in the mine were familiar with the dangers connected to the use of explosives. Former Sterling Hill miner Clarence Case tells of losing a friend.

Figure 42. Fire inspector and team lay hose during a plant fire drill. (Photo: *Zinc Magazine;* Courtesy of Ogdensburg Historical Society)

The machine runner drills all of the holes. He gets in his dynamite, puts the sticks of dynamite in the holes, and then he does his blasting. That can be dangerous if you're not careful. I lost a good friend of mine because he miscalculated the time that we required to leave the explosion area. He lost his life. Once you start loading the holes in the raise, you need approximately three minutes — once you start lighting them — to leave the area to get far enough away where the explosion won't bother you. He miscalculated. He lit all the holes and he just didn't get away in time. And that was it!

Steve Misiur, who worked with miner John Kolic while drilling new tunnels for the Sterling Hill Mining Museum complex, describes another problem in working with dynamite.

Handling the powder, particularly if it was dynamite, that could be detrimental. We use a non-nitroglycerin explosive here, but we also use a nitroglycerin explosive, that's dynamite. Dynamite is nothing more than sawdust with nitroglycerin mixed in with that sawdust, in different strengths anywhere from 20 to 60 percent strength. Now, the dynamite comes in sticks about seven, eight inches long, wrapped in a very stiff cardboard tube. Two problems with that dynamite. One, you could not allow it to age. If you let it sit and let it age, what happens is the nitroglycerin sweats. It sweats right out of the cardboard and it crystallizes. It gets so sensitive that the least speck will set it right off. You don't want to use that. But the other problem is the nitroglycerin. You see, if you have a stick of dynamite in your bare hand, the nitroglycerin absorbs right through your skin causing your blood vessels to dilate, to expand, particularly the ones in your brain. Those situations produce what I'll call the infamous powder headaches. Powder headaches, because what happens, the blood vessels in your brain expand so rapidly, so fast, that the blood pressure literally just soars. That sensation, I can describe to you. It was as if somebody was hitting you on the head with a claw hammer. It was a pulsing headache that lasted anywhere from six to eight hours.

We used dynamite for several reasons. Because it has a better breaking efficiency. The velocity of the explosion is such that it produces small pieces of muck. You want to make the smallest pieces of muck because it's easier to haul the stuff out. But if you use dynamite, you have the other problems. So to get around that, they used the non-nitroglycerin base stuff. The velocity of that stuff is different so that you've got bigger pieces of muck. It's harder to take that stuff out, but then you have the advantage of not getting powder headaches. There is no cure for these powder headaches. They vary from one miner to the next miner. Some poor slobs had them for two to three days. Very, very detrimental to use dynamite.

John Kolic worked in Sterling Hill until it closed and he, like Steve Misiur, recognizes that dynamite presented problems.

The biggest technological change that affected us as miners during the time I was there probably would be the explosives. Dupont stopped making dynamite. Relatively speaking, it's a hazardous substance to use. They developed a variety of what they called water gel explosives, which were much safer. Much safer. Very difficult to set them off accidentally. So that was a good change.

The rock that you're leaving behind around you and overhead has to be properly secured so it won't fall on you. That's the most common accident in mining, falling rock. I've seen big rocks fall down. We had many cases. I never had a lost time accident because of rock falling on me, but others have. Sometimes it's carelessness or inexperience. Poor judgment. To make that rock safe, you have to do something to it. Install rock bolts, support it with timber. In selecting how you use those tools, you have to use judgment about the ground conditions, about the strength of the material you are using. If you make a misjudgment, what you do maybe won't work.

Tom Sliker, who worked at Sterling Hill for thirty-seven years and saw many technological changes, agrees with John Kolic that good judgment is essential to the safety of a miner. Tom says:

As far as mining is concerned, you have to [put] emphasis on the safety at all times. What I learned the first couple months I was there is that you have to take care of yourself.

You don't wait for someone to do your safety for you. You have to do it yourself. We didn't have that many accidents, but most of the accidents that I saw were caused by not doing the right thing at the right time. Carelessness. They had a book of safety rules. You had to sign that, that you did read it. Your rules were to watch for loose, to take care of yourself. Being on time. Be alert. Wear your goggles at all times. Watch for your loose [rocks]. That was the main thing. The last thing in that little book was NO HORSEPLAY! I can remember right now reading it, the last thing in that book, NO HORSEPLAY!

A good friend of mine got killed by horseplay. A boy from Franklin and his partner. They were buddies, good buddies. Young. And they were always horseplaying with one another. Especially at quitting time. They would be on the station waiting for the cage to come and take us up. It's just like being in New York in the subway, waiting in the station for a train to pick you up. It didn't happen in this mine. It happened in Mount Hope Mine. This one night, they were down in the mine and this fellow came in with a new lunch pail. All over a lunch pail, he lost his life. They were way down in the bottom one night. They used to use a skip to go up and down to that level. It didn't come from the surface. It just come from say twenty-one hundred up to maybe seventeen-fifty. So they used to get on this skip and the hoist man would bring them up. The incline was about fifteen degrees, and the chutes come out, to hoist the ore up there was only about six inches clearance. So this one part here was narrow. He says, "Where is my lunch pail?" Somebody had hid it. He was looking for it and he put his head out over the side here and the skip took off. It just rolled him through this little narrow. Killed him. All just because of having fun.

In spite of the Zinc Company's very good safety record, there were some terrible fatalities in the Franklin and Sterling Hill mines. While falling rock was most often cited as the cause of accidents, the man cage or shaft elevator was a particularly dangerous place and, as a result, was often mentioned in warnings from the company. In the May 1939, issue of *Safety Age*, there was a description of a scraped arm accident that occurred while the miner was on the man cage. The evaluation of that accident warns:

> *If men keep their arms inside the cage there is no*
> *danger of injury from being struck by the shaft*
> *timbers. However, if any part of their arms or hands*
> *is allowed to protrude over the side of the cage, there*
> *is the possibility of a serious or perhaps a fatal accident,*
> *since there is very little clearance between the side of*
> *the man cage and the shaft timbers.*

Several narrators, including Harry Romaine, Tom Sliker, and Steve Sanford, told of the many dangers and specific accidents involving the man cage. Harry Romaine, a former miner, speaks about safety and recalls riding on the man cage at Sterling Hill. When Harry says that a miner did not have to go anyplace that he did not consider safe, he is referring to provisions contained in the ratified 1974 United Mine Workers contract, which stipulates that any miner can refuse to work in a work place that he considers unsafe "without fear of reprimand or dismissal." This protection was a major victory for workers.[2]

> We put thirty-two men on it. Yup. Every once in awhile, your hat would say *click-click* on a beam. Somebody pushed you out too far. If there was a guy in back and a big heavy guy in front of you, you couldn't get all the way in. [laughter] If you got in by a big person, he took up room for three. You learned to hang tight. If you heard your hat click or felt it, well, you got your head back, or it would hit one of the beams going down.
> There was a lot of emphasis by the company on safety. At first. But, then after, they were kind of lax in safety as things went on. They wanted too much for too little. They would want you to work in a place where it wasn't properly timbered. In other words, you were taking a chance going in there. As it was, everything panned out all right. We did have a safety man there until he passed away. But he was a safety man! He wouldn't send nobody anyplace that he wouldn't go himself! Syd Hall. He was the safety engineer. Then there was another one, Joe Weeks. Joe was a safety man under Syd Hall, and they were good. And they done a good job. But sometimes you got a boss just wanted to make a name for himself and he'd want to send you here, but you'd just say, "Nope, Syd wouldn't want us to go in there." He had to accept that. That's right. They could not send you any

place that the safety engineer had not okayed. Everything had to be under the safety boss.

One of the worst fatalities involving a man cage was related by former Sterling Hill miner, Steve Stanford. Steve had been offered a job as hoist engineer, which he refused since he did not like the responsibility of working as either the hoist engineer or the cage man. Steve describes the man cage and the work involved.

Have you seen a man cage? It looks like a set of steps on rails. It goes up and down the shaft. It carries men and supplies up and down. It was like a set of stairs and there were three men per stair and you were all packed in like a can of sardines. Tall guys had to tuck in because there's no protection about. You'd bang your head off the eye beams as you went past. As a matter of fact, one day I was in the cage and these two guys were joking around and I got pushed as the cage was moving. I got a beam right in the top of the head. I was real lucky it didn't break my neck. I was kind of stiff for a couple of days. But that was one of the things you had to look out for, especially if you were tall. It wasn't an enclosed cage. You actually were riding open exposed to the beams that supported the shaft. It sounded like a railroad coming through.

The regular cage man, he was the guy that took the orders for equipment, who to pick up and so on. Well, he often went on a five-week vacation in the winter. He was a pro there. And I often got stuck being cage man while he was gone. Oh boy! All day long. The bosses had to get around and men needed equipment. When you were in the station and the phone rang, you had to get off the cage, answer the phone, and they'd say, "Hey, I need a jackleg drill on eighteen-fifty." So, you'd make a mental note, "eighteen-fifty needs a jackleg." And, you'd go up and down the shaft all day long answering. And if a boss needed to be picked up, he'd ring the bells. There was a code for each of the levels. If you heard a one followed by a quick four, that would be the four-thirty level. You'd have to tell the hoist engineer to take the cage to the four-thirty level. You'd hear, say, three-five, and that would mean you would have to go down to fifteen hundred level and see what they wanted. All kinds of stuff. You're going up and down the shaft all day long, trying to remember what level you're on, what that level wanted. It was mentally wearing. After a day, I'd get home and I'd think, "I can forget about that doggoned cage for awhile." I'd go to sleep and I'd dream all night long. Up and down the shaft, ding, ding, ding, ding, ding, all night long!

The way a cage worked, before they had fail-safe, there were actually two cages together. One in each shaft and when one went down, one went up. Supposed to be in balance. You had to do that in order to be able to control it accurately. At Sterling Hill, they were balanced. I don't know the mechanics behind it but it was much easier when one was going down and the other's going up. One day in Sterling Hill, they took it out of balance. They were loading a single cage with men on it and it got away. It went screaming down the shaft. A man who was over six feet, when they took him out, he was about five foot eight and his leg bones were driven through the soles of his boots. It was horrible, it took several days to clear up that mess. But you had to accept the danger. It was part of being a miner. You had to accept the fact that someday you might be coming out feet first. Most of them laughed it off. I had a number of really close calls. You give a shrug and go on with your work. Most miners didn't spend a lot of time talking about it, but almost everyone had close calls.

One day, I was in a place that was well known for its bad ground. I was there with a boss. He was just a top miner. See, somebody fired the last shift, blasting, and the ground wasn't secured. Just a pile of muck there about fifteen feet deep. This was unfired ground I was standing under. The other guy climbed up on top of the muck pile. He was prying around up there and that was absolutely against the rules. Even though he was a super miner, he was up there with a scaling bar tapping rocks on top of the muck pile, which hadn't been secured. You're supposed to scale and roof bolt before you get under bad ground. He was up there prying around on top of the muck pile and all of a sudden, he got covered. It all went SZHANG! And I don't see Bob. Just his feet sticking out of the muck pile. I knew it was bad. I went back there expecting to see the worst! I was quite sure he was a dead man. It looked like the whole roof was falling in. But he was okay. He was up there on top of the muck pile laughing at me. He said, "I didn't know you could move that fast, Steve." I wouldn't have done it under any circumstances. He was an exceptionally good miner, but he was not supposed to be where he was.

Every Friday, we had a safety meeting. Every section underground had to have a safety meeting. It went on for forty-five minutes or so. Any lost time accidents were discussed in detail, trying to figure out how it could have been avoided. Why that guy was so dumb as to do something stupid like that. That kind of stuff. The shift boss conducted them. If a section had a three-month period without any lost time accidents, everybody got a safety ham. Something that would give you a little reason to avoid lost time accidents. If everybody in the crew got a safety ham, maybe the guy who lost the safety ham in the other section, they'd look down on him. You know, give him a hard time about it. It was a way they felt of promoting safe working by the miners. You didn't get any money, but your crew would get some sort of reward at the end of a three-month period. But, if you had been in an accident and could not work for a few days, your whole crew lost the safety ham. You had to listen to them for the next three months yelling at you.

Bernie Kozykowski, who worked as a miner at Sterling Hill for several years, felt that the danger inherent in mining was also part of what made the work so interesting. The shared acceptance of that risk was part of the close working relationship among the miners. As Bernie put it:

> We take care of each other. I mean, you've got to. You have to understand that when you're working underground, it's dangerous. Mining is a hazardous occupation. You depend upon your relationship with your fellow workers. You fight amongst yourselves like crazy, like a bunch of high school kids. It's really funny. I mean, we used to really scrap all the time, but, at the same time, you could always depend upon one another if you needed help. It was very close, very, very close. From the mine superintendent right on down. You had to. When we worked underground in the mine — you have to envision yourself being deep in the bowels of the earth, sometimes as much as half a mile down and in an environment that is anything but friendly, where the air is foul, and, typically, damp. Very often full of all sorts of noxious gases where you're working. To

Figure 43. Today, the Mine Safety and Health Administration offers safety courses at its Academy and training resources on its website. (Image: Courtesy of the National Mine Health & Safety Academy)

give you an example, there were times when we would be working in production drilling and blasting the ore, and, more often than not, we would be drilling with a hand-held machine called a jackleg, which is a jack hammer with a piston on it. We would be working with this hand-held drill, and in order to power it, you have an air hose and you have a water hose to keep the dust down, and this thing is working under great pressure, a hundred and seventy-five pounds of air. You're drilling and hammering and you have a steel on the end of the machine that is hammering its way into the rock and these things would snap off on occasion. And, very often when you are doing this kind of work, you are standing on rock that has been broken and is very sharp and jagged. The air is foul and if the steel would snap, the machine would whip around and grab you and throw you down, and would be beating on you before you got to shut it down. You would be all nicked up and cut up and bleeding. You could barely breathe and in frustration, you're crying. But it's not a mournful tear, it's just frustration, and you talk about blood, sweat, and tears. People in the real world have no sense of what some people and some industries do to make their lives comfortable. I went to the hospital five times with injuries, but I never lost time. You're always getting beaten up in the mine. It's part of the game. You toughen yourself. The difficult and dangerous conditions, that's part of the appeal. Yeah. There's no doubt about it.

Even though, as Bernie Kozykowski said, "Mining is a hazardous occupation," it is far less so today than it was in the past. Now, enforcement of all laws and inspection of all mines are the responsibilities of the Mine Safety and Health Administration, a branch of the Department of Labor. In addition to the strict enforcement of safety codes, mandatory training of miners and safer machinery have made American mines considerably more safe than they were even ten years ago. Today, the mining industry has a lower rate of injuries and illnesses per one hundred employees than does agriculture, construction, or retail trades.

1. Safety Age Booklet, New Jersey Zinc Company, 1939.

2. US President's Commission on Coal, *The American Coal Miner*, 33.

3. Levy, *Struggle and Lose, Struggle and Win: The United Mine Workers*, 8.

4. US President's Commission on Coal. *The American Coal Miner*, 33.

Anomalies, Oddities, Pranks, and Ghosts

For decades, a concrete stanchion used as a base for the hoist that drew crude ore from the Parker Shaft stood unused. When the Kiwanis Club of Franklin erected the Franklin Mineral Museum, that stanchion remained as a worthless obstruction on the lawn of the museum. It caused some interest because the sand used in the concrete was made of the crushed ore and at night, with a portable black light, it would fluoresce with the brilliant hues of the Franklin minerals. Today, the stanchion serves an honored purpose. It is the base for a wood-carved statue, memorializing the men who worked the Franklin and Ogdensburg mines.

New Jersey Herald, May 1972

Unquestionably the most spectacular thing about the Franklin and Sterling Hill mines was the rare ore body itself. Not only were these great mineral deposits extraordinary because of their inherent special properties, but also because they remained unsurpassed, in the quantity and quality of these properties, by mineral deposits found in other parts of the world. During the years that the Franklin and Sterling Hill mines were operated, more than thirty million tons of ore were extracted. The most important material taken from these mines was, of course, the very rich zinc ore, but many other minerals were also removed. In fact, about ten percent of all known mineral species in the world can be found in this area, and several of those species cannot be found in any other place on Earth. Seventy minerals from the Sterling Hill and Franklin mines fluoresce, an unusual property which is now being featured in the Smithsonian Institution's two-story mine re-creation. Visitors to the Smithsonian exhibit are surrounded by the brilliant reds, dazzling greens, and vivid oranges of a fluorescing ore vein from Sterling Hill that is believed to be one billion years old. According to the Smithsonian, the intricate natural process that formed the unique fluorescent minerals from Franklin and Sterling Hill remains a scientific mystery.

Figure 44. With a portable black light, the concrete stanchion used as a base for the statue fluoresces with the brilliant hues of the Franklin minerals. (Photo: Carrie Papa)

Beyond the scientific and popular allure of mineral fluorescence, however, there were other peculiar thoughts about and happenings in the mines. The incredible depths underground, the miles of tunnels, the sometimes total darkness, and the omnipresent danger merged to create an atmosphere that was conducive to strange thoughts, spooky stories, and rowdy behavior among the miners. The tensions created by the hard work and surroundings often were released at the end of the day as the miners joked and played tricks on one another.

It was a natural thing for miners to believe in ghosts. The hundreds of miles of drifts and tunnels, the frightening thought of being so deep in the earth, the eerie sounds and strange smells within the mine, the appalling dark, and the unrecovered bodies of miners buried underground all contributed to the sense of the unnatural. Superstitions, ghosts, and hauntings were a familiar part of mining, not only in America but also throughout the world. For example, the superstition that any woman entering a mine would bring bad luck was universally accepted.

The Cornish brought a great many superstitions and legends to the mines. The "Cousin Jacks" believed that if a candle went out, the miner should leave the mine immediately. This myth is based in the fact that if a candle is flickering or goes out it could mean that deadly gasses are in the air. The Cornish also believed that mine rats must never be killed. As with canaries, any distress to the rats could alert miners to the presence of poisonous gases in the area. Also, the Cornish prohibited whistling in the mine during the night shift because sharp sounds were thought to cause vibrations that could result in a cave-in.

In the face of all the irregular dangers that surrounded a miner, it is easy to understand why the hours from eleven at night until seven in the morning were called the "Graveyard Shift."

Nearly all of the narrators featured in this book commented with pride on the distinct properties of the Franklin and Sterling Hill ore deposits. The remarks of Bernie Kozykowski are typical.

> I remember coming here [Franklin] as a boy when Uncle Eekie was still working in the mine. I was nine years old. My first recollection of the minerals in Franklin was the fluorescent aspect of them. That characteristic. You put them under an ultraviolet light and they sort of glow in the dark. It's not unique to Franklin. At the same time, there's no place

else on the planet where you have such a proliferation of fluorescent mineral species. I think that's what makes Franklin very, very special.Somewhere around thirty years ago, a group of people from Franklin was able to get a state assembly declaration that proclaimed that Franklin is the *Fluorescent Mineral Capital of the World.* Appropriately so. I've written several articles relating to the mineralogy and the fluorescence and so forth, and I wouldn't disagree with that. Typically, in most places where you do have some sort of mineralogy occurring, if you encounter two or three different

Figure 45. Evan Jones (Uncle Eekie) and a fellow miner drill ore in a Franklin pillar. (Photo: *Zinc Magazine;* Courtesy of Ogdensburg Historical Society)

mineral species that might respond to ultraviolet light, you have a great many. Here in Franklin, there are well over fifty. So it is rather phenomenal. There isn't any place like it on the planet that even comes remotely close. Scientists have identified more mineral species occurring here in Franklin and Ogdensburg than any other place on the planet. The mineralogy of Franklin is exponential in its impact as opposed to being very basic and common. It's the uniqueness of it. The strange thing about it is, along with this, is the richness. You've got to consider these two zinc deposits are the two richest zinc deposits in the world. Most zinc mines pursue ores that will run anywhere from six to seven percent zinc in what's considered a good deposit. This is after concentration. At Franklin and Sterling Hill, they started out with twenty-two percent. Can you imagine the wealth that's there? The other thing is that this had to be Mother Nature's consummate stew pot because in chemical terms there is an awful lot that occurred here. I think that has caused more things, more scientific papers regarding the mineralogy and geology in technical terms, to be written about Franklin than any other place in the world.

Robert Metsger, who is one of the most respected scientific writers on the Franklin and Sterling Hill minerals, tells about his work as a geologist in these mines. Bob also reports on the Sterling Hill mine being designated as one of the first two sites the worldwide network of seismograph stations set up by the United States National Oceanic and Atmospheric Administration.

Franklin is a very famous place. To a geologist, particularly mineralogists, it's famous all over the world. When I started in Franklin, we had a chief geologist, Alan Pinger, and Jack Baum was there. What was being done, we were mapping the entire Wallkill Valley, actually. Everything from south of Andover, Byram actually, north to Pine Island, Mounts Adam and Eve. I had a very small part in that because I came in fairly late, but the department did a geological map for that whole area. That was published in the *Geological Society of America Bulletin* and also used in the US Geological Survey Map for this area. I mean, it's the definitive geologic map of the area. While I had some part in that, my chief job was working in Sterling Hill. I was the geologist for Sterling Hill because Sterling Hill was involved in a lot of geophysical exploration in the area. Seismic and electromagnetic. I used to go down every day and track the operation as it progressed and find out what the genesis of the ore body was or how it got there. We think we have a very good idea. It's not what it was thought to be back when I first started. I have published on that and I still intend to do some more publishing.

I brought a copy of a seismology record so that you would know what I was talking about. These quartz tubes record movement between the piers. Each one of these lines is an hour and shows the effect of the moon flowing over the crust of the earth from day to day. At different times of the month, different phases of the moon, this might flatten out. Sometimes it gets very low. Now, this section here, this is the Chilean earthquake back in 1960. Here's when the earthquake occurred. These surface waves going around the earth, those waves went through the mine at Ogdensburg over thirty times in three days. That record is the first time the free period of oscillation of the earth was ever measured was ever recorded. That's when the earth was ringing like a bell or pulsating like a heart. The actual date of that earthquake was May 22, 1960. I did quite a lot of work with that and published, co-authored, quite a number of papers on that. That became a very important station. It was on the eighteen-fifty level of the mine, which is about twelve hundred feet below sea level. It's quite a famous record now. It's published all over and used for a lot of research. Because that was so successful, that particular site was selected by what was then the US Coast and Geodetic Survey. They were setting up a worldwide network of seismograph stations that were everywhere in the world except behind the Iron Curtain. The purpose was to monitor atomic and hydrogen explosions. The mine here, the Ogdensburg mine, and the Western Observatory up outside Boston were the first two stations on that network. It was a part of Columbia University. Actually, you might say a subsidiary of the Lamont Geological Observatory.

In addition to important seismographic records, other narrators report on strange temperatures and atmospheric conditions in the mine. For instance, Sterling Hill miner Harry Romaine re-members ice forming in the mine in the middle of the summer.

> Above the five hundred level was worked from the top. From the surface. That had what they called a three-forty level, a one-eighty level. The one-eighty level, there used to be icicles hanging there in July. It set right on an ice canyon, where the water would drip and freeze. That would be yet in July.

Franklin geologist John Baum speaks of the pressure in the mine and also remembers the cool temperatures.

> We went down to the eighteen-fifty level of the Sterling mine and we put down a hole 5,200 feet deep. Now, that's a deep hole for the kind of drilling we were doing. It was called diamond drilling. It recovers a drill core so that you can look at it and actually see the rock that's down there. We got all that drill core from way down there. We were told by the superintendent, "Don't find another ore body a mile deep." Because the kind of rock we have here tends to close in a mile down. Pressures are too great. Anyhow, we were down considerable in our mine. We have a drill hole next to the Franklin Pond here and it's six thousand feet deep. Now, the temperatures as you go down in the earth get hotter and hotter and hotter. The one down Sterling, the deep hole there was ninety-six degrees at the bottom. It would be nice if you could pipe that up, you could use it in the winter. We're fifty-three degrees here. That's what the temperature is in the mine. Except for the fact that you've got a certain amount of circulation from the surface and you're introducing compressed air and so on. The men give off body heat. But, in the mine, it's generally cool, so we wore long johns. It's an interesting experience to come up out of the mine on the fifth of July wearing long johns and that heat hits you. Bingo, you droop. [laughter]

Discussing the roughhousing in the mine, Sterling Hill miner John Kolic says:

> In the morning, the miners would come in kinda slow, not much excitement, get their mine clothes on like they're

not really looking forward to going in there. But, at the end of the shift, when they came out, they were exuberant. Like a bunch of kindergarten kids in a schoolyard screaming at each other and knocking each other down. They ran out of the mine like wild Indians. Ran down to the change house. Ran to be the first to get into the shower. Would be yelling insults at each other and, in general, having a good time. There was more excitement at the end of the day than there was at the beginning of the day.

Some of the cage men just did their jobs. Others went out of their way to cause mischief. I saw one case where myself and another miner were working on a job away from the shaft. There was a wooden table at the shaft station where we had our lunch boxes. After lunch, we went back to work. When we come out at the end of the shift, someone had nailed my buddy's lunch box to the table from the inside. We thought the cage man had done that during a slack moment when he had nothing else to do. That was a pretty good trick. And it was interesting too because that's not the kind of thing you discover right away. You come out and the cage isn't there yet. You go over and you sit down and you try to move your lunch box and it doesn't move. You open it up and you say, "Boy, look at that." What happened was the guy was in a hurry. He grabbed his lunch pail to take off and it wouldn't move! We fixed that cage man. Sprayed him with water pretty good. He was coming down on the cage and we had these three-inch plastic hoses for the back fill. We had one about fifty feet long curled up sitting on the station because they had to do a back-filling job on that level. From the water fountain, we filled that fifty-foot length of three-inch hose with water. Then there's a clamp where you could put an air hose from the compressed air line on to the end of the pipe with a valve and hang the other end of the hose into the shaft. You could hear the cage coming down. It makes a lot of noise, rattles and what not. When it sounded like it was about to get to our level, somebody turned the air on and blew that fifty feet of water right out at the shaft and it just so happened that the guy arrived at the right time and he was drenched. That was a pretty good trick.

Along with several other odd events, former miner and mine superintendent Tom Sliker tells of an incident involving John Kolic after the Sterling Hill Mine had been closed. Tom also recollects the constant ambiance of danger and darkness, ghost stories, his own nightmares, and the strange feelings some miners expressed.

John Kolic used to test hole for me when I was running the mine. He was a great mineral man. He knows where all the minerals in the mine are and what they are. After the mine shut down, he and another guy used to go down there and mine. He got an air machine to make air. They'd go down there and drill for these specimens, blast it, and then the next day, he'd go down there and bring up all these specimens. The Hauck boys owned the mine then. Anyway, one morning, he was setting outside and was going to go down, and all at once the whole place started to shake. "What in the world was that?" Then they were going to go in and it happened again. It was a tidal wave! Now the mine had water from the bottom up to the five hundred level. The south side of the mine was mined years ago and that section was filled with sand. They used to have a mill that separated the ore here. All the sand came from it. Instead of buying dirt, they put the sand down in here. And that area was so dry, and it was warm. A lot of people used to go back there just to eat because it was nice and warm at lunchtime. You could get a hand full of sand and hold it like that [cupping hand]. It's like sugar. You can't keep it in your hand. That's how bad it was. Well, these pillars started to collapse. Pieces as big as this house would fall out leaving big holes and voids. Some of these stopes would run empty and go down some of the others. Well, what happened when that mine got filled, got this water in it, all this must have started moving over here and it went down. Like I say, you had two shafts. You had the old shaft and the new shaft. Well, when that went down, it pushed that water and it came up to the old shaft to the five hundred level. So powerful and there was a boulder in the rush. It must have weighed four tons. It just pushed that boulder right up through the drift. If they had been down there that day, it would have killed them both. The pressure was so great. That whole section of that mine must have just blew at once. That happened a couple years ago. Just before he was going down. He would never have got out.

You don't think about things like that. Well, sometimes you do. You know, I do more mining since I've been on pension than I did when I was working there. Every night I would dream about mining. Mining is a very dangerous job. All those things that I thought about that could be dangerous, I dream about now. Now when I dream about mining, I'm always lost. Can't get myself out. I'm always there trying to get out and there's always a lot of water.

You don't know what darkness is unless you go down in a mine. It's so dark that you don't know whether you're standing or laying down. It's just so dark, you don't know nothing. When I first went in the mine, we was wearing carbide lights. Around three-thirty everybody would be blasting, blasting their round off for the day. When that blast went off, the concussion went through the air system, out goes your light. Everybody's light would go out. It was pitch black! I remember the first man I worked with on a machine was Wulfus from Ogdensburg. He's the one taught me how to drill a raise. He was a Dutchman. Real big guy. Bony too. Real rough character too. He knew his stuff though. We're about seventy feet up in this square hold. No ways to go except down. We used to drill our round, load it with powder. In those days we had burning fuses, like you see the cowboys with them lighting them. Well, we're up there and we're lighting them and his carbide light goes out because if you get in front when you light the fuse, it blows. So he took mine. And mine went out. There we are up there and pitch black. But we always had a chain going from where we were working down the raise. We just got hold of that chain and got down as fast as we could because these fuses are burning. Somehow that frightened me, I guess, because now I can't go to bed nights without a night light on.

We had one man came to work there when Jim Onder was boss on the eighteen-fifty level in the north ore body section. He went down and he saw this guy down there walking around like this, looking scared, you know. He said, "What's the matter with you, Buddy?" The guy said, "I don't know, it feels like all the sides are all coming in, you know." So Jim took him off. Laid him off right there.

What do I know about the ghost of Bicycle Pete? Well, this man Bicycle Pete — he got killed in the mine. In the north ore body. You know, some people are loners, I guess. What I knew about him, I just knew who he was. He never had, to my knowledge, any friends. I never saw him deal with anybody in the company. He was just always by himself. Somehow, he fell down an open hole and got killed.

$\sim$

It was hardly possible to be a miner without knowing of deaths in the mine and believing in the possibility of the ghosts of those miners. Many of the narrators were familiar with stories of dead miners whose spirits continued to roam the black corridors.

Steve Sanford, who worked at Sterling Hill in the 1970s, gives additional details on the ghost of Bicycle Pete, tells other ghost stories, and discusses drug use in the mine.

> [Bicycle Pete] was alive just before I got into the mine, but he fell down a raise in the north ore body, four hundred and fifty feet. He used to bicycle from Franklin to Ogdensburg every day. He was a foreign-born fellow. And he could be seen haunting the north ore body every now and then. One fellow I know — in fact his grandmother told me this story — he had told his grandmother how he was sitting there goofing off one day in the north ore body, and he had his light out. He was sitting in a drift just twiddling his thumbs waiting for quitting time, and he saw a light coming down the drift. He thought, "Oh Lord, here comes the boss." And the light came down the drift exactly like a man would be walking. Then turned to the left and disappeared like he was going down a side drift. The miner's name was Mickey; that's what his grandmother called him. He kept his light off and snuck up and went to look down the drift to see where the light had gone to. He got there and there was no cross drift there! The light had come down the drift, turned, and disappeared. He decided it was Bicycle Pete walking around.
>
> The other standard story is Luke the Spook. He was in the middle section and he lit one too many burning fuses. He was vaporized. They couldn't find all of him so they had to close off that part of the ore body and not mine it. Theoretically, Luke the Spook is still in place somewhere in the Sterling Mine awaiting the Resurrection. When I first got there, they told me — we'd come out of the stope for our breaks and sit on the platform on the level leading down into the place. They told me, after I was done with the coffee break, I had to leave a cup of coffee or something for Luke the Spook to come and take.
>
> One day I'd been tramming from a place on eleven hundred. We had lunch and I went looking in the back for specimens with my buddy. We walked into a dead end part of the East Limb. It hadn't had any work done there for years

and years and years. It was all blocked off, this dead end. We were looking for rocks and we were walking along there not saying much. I noticed that I could hear a couple of men talking. I couldn't make out what they were saying, but very distinctly heard men talking. I thought, "Oh, oh. There's no possibility of hearing anybody in this drift because there's no access to it, so I'm not going to tell Jones that I hear people talking." Then Jones said to me, "Hey, do you hear a couple guys talking?" We turned and left quickly! It sounded exactly like two old men talking about the good old days.

I talked with a fellow who worked in the last days at the Ringwood iron mines after the Second World War. Some of the guys at Ringwood had been there for generations. One day a guy came down and said, "Well, I saw Uncle Milt up in twenty-one hundred stope." He didn't think anything about it until somebody says, "Uncle Milt's been dead for years." The mine was the sort of place that encouraged these things.

In the seventies when I worked there, just past the psychedelic sixties, there was some drug use underground. Some guys were caught using dope and they left in a hurry. There are some stories in that too. One guy I remember was doing LSD one night. He was supposed to be tramming. He had filled his ore cars and he was motoring down a drift. He was carrying ore from a working place to a collection spot. He was chugging along. All of a sudden, he looks at the drift ahead of him, and he just went *eeiyeike* and stopped. He was facing a blank face of rock all of a sudden. He couldn't do anything, but just sit there and wait for it to open back up again. When the boss, Bob Morris, caught him, he says, "What are you doing here, Maaco?" "Oh, I'm waiting for the drift to open up, sir." "What do you mean? "Ahah, well, ah . . . ." And the boss says, "Out of here!" He was gone!

Another of the guys I worked with for awhile liked his pot, his marijuana. I think that's what he was doing one night. He went off by himself for awhile. When the cage came to go home, he wasn't on it. So they sent somebody back in to look for him. He was in an old part of the mine on ten hundred, waiting for the monsters to go away. He'd been walking alone. The mine could be a very spooky place, especially when you're alone. He was walking along there enjoying his marijuana high and he heard, brrurp, brrurp, brrurp, and he was sure it was a monster. He just stood there and he waited for about an hour till they came looking for him. It was an old air pipe leaking bubbles in a puddle, but he was sure it was a monster. He wasn't fired for that, but it was quite obvious why he was still there waiting. It was not really smart of you to work in the mine impaired. The mine could be a very forgiving mistress, but if you kept doing something dumb, why you could get slammed for it. I know one fellow who was stoned and fell down a raise, and he couldn't work another day in his life. With one or two exceptions, drinking was almost always after hours.

Accidents involving drugs or alcohol were very rare. Safety reasons alone were enough to inhibit drinking while working. However, as two men report, having a drink before or after work was fairly common. Although nearly everyone agreed that the miners were quite heavy drinkers, Bill May claims that there were very few incidents involving intoxication. According to Bill:

Figure 46. Narrator Steve Sanford says, "The mine could be a very spooky place, especially when you're alone." (Photo: Carrie Papa)

These fellows knew that they had to go to work and they knew you better not be drunk on the job. But Snyder's Hotel was open even in the morning. The men would go in there and have their shot and beer before they went to work. That wouldn't bother them. All a bunch of tough cookies.

I can remember when I went into the Zinc Company after the store closed, I worked in what they called the shipping gang. We filled ore cars with the ore. We had this tough little Hungarian fellow there, and he'd bring in this old quart mason jar, have the zinc lid on it, and he's got hot peppers in there. "Aaahh, have one, Billy," I had put that just to touch my lips and it burned all day. Anyway, after it was gone, the juice that's left in there, he'd drink it. My God. His name was Matty Putuchi. Tough, little Hungarian guy. Boy, he was about five foot tall, but tough as nails. All fine workers. No, a shot and beer in the morning wouldn't bother them. Not these guys.

Sterling Hill geologist Bob Svecz confirms that having a drink on the way to work was a common practice with many of the miners. Additionally, Bob mentions the number of bars in town, and relates several incidents and jokes told to him by his father, who worked in the Franklin Mine.

In Franklin, if you walk down Main Street, how many bars were there? How many liquor stores? It was a cheap source of entertainment. A lot of the miners at breakfast would have a shot and a beer before they went to work.

If you had a late night out on the town, you might doze off during lunch. One guy, he was a sound sleeper and during lunch hour, he had fallen asleep on some of the wooden planks they had used in the mine. Smiling, they nailed him down to the plank with his clothes.

My father told me about this one guy who didn't wear a belt. He used a fuse line to tie his pants up. At lunch time he took a nap. He was a sound sleeper too, and the guys lit the fuse on him. They weren't very hot. You could throw them in water and they're not going to go out. They lit the fuse on the guy and he had to untie his belt to get the fuse off.

In the first aid boxes, they had stretchers in them. Some of the guys used to take a blanket, wrap it up into a pillow, and take a nap in the first aid box that had the stretchers in them.

You could lay down in the stretcher and be real comfortable. And it was warm. You had the two light bulbs on. They had a lock hasp on them, but they never used them because the first aid equipment is supposed to be accessible. But the only time I've seen them used, this guy would be in there taking a nap. His buddy would come along, slip that hasp over, maybe put a nail or a bolt through it, the guy's stuck in there. At the end of the shift, when they came to pick him up, he wasn't standing there out in the shaft station. The boss had to come over and let the guy out. The boss was annoyed about it. He didn't get fired because he couldn't lock himself in there, but the boss knew the guy had been in there since lunch time and now it was quitting time, and he hadn't done any work.

Franklin had rats, and if you found a dead rat in the morning, you put it in the miner's lunch bucket. Of course, this is after lunch and he wouldn't check it and take it home. His wife would open it and there would be the rat in the lunch box.

I've heard stories where guys would be sitting around for lunch, and for a joke, another guy would get a stick of powder, dump the powder out of it, fill it back with sand, get a burning fuse — of course no cap on the end of it — light the fuse and throw it in where a group of guys were. You couldn't tell if it was real or not. The fuse would be lit and you didn't know if that was a real stick of dynamite.

In spite of the physically hard work, it seems that most of the miners, with their jokes, tall tales, and hazing managed to get the most enjoyment out of the workday that they could. On the other hand, Steve Misiur, who worked on the Edison Tunnel in the mine museum after Sterling Hill closed, reflects on the physical demands of being a miner. In fact, Steve connects the hard work to the isolation a miner feels. The intensity of the work, the isolation, the ever-present danger would combine and contribute to the development of superstitions and ghost stories connected with most mines.

The physical nature of the job is wearing. The other bad thing about the job is that it's a very isolated, very, very solitary job. It's a very solitary profession. You're working with someone, but because of the nature of the job, you can't communicate like you would sitting down as we are right now. You couldn't do it because you're working for almost every single minute. The only time you could have a conversation is on your lunch break, which is only thirty minutes long. Also, you're so physically worn out that just sitting there is physically grinding. You just don't have the motivation to strike up a conversation with anybody. So that's one thing, this solitary nature of the work. You can be standing next to a miner all day long and yet not sense that they're there. Also being in that dark environment, in a cramped and darkened manner like that kind of deprives you of your normal sense of connection to the regular world around you. You're in an area of the mine in a dark environment that smells, that's cold and wet, and noisy to no end. Can't even hear yourself think. That's probably one of the worst aspects of the job.

Even if conditions in the mine were harsh, and the work of the miner difficult, most of the narrators were pleased to have been part of the New Jersey Zinc Company and were proud of the work they did. Steve Sanford speaks for many miners everywhere when he says:

> Not everyone could put up with the working conditions. It could be quite tough. But you could feel a little pride in being able to do something not everyone could do.

Rock Hounds and Pebble Pups

Franklin has been called the Mecca of mineralogists. From the time when Dr. Fowler first brought this field to the attention of scientists up to the present time, more than one hundred minerals have been discovered, investigated and named. This is a greater number than has been taken from any mining district in the world. Many of these minerals are found only in this locality. Many of the minerals are massive in structure; others are in the form of crystals, sometimes transparent, of symmetrical structure and of beautiful color.

Historical Notes of the Iron and Zinc Mining Industry
Sussex County, New Jersey

About four thousand known minerals have been found on Earth, and some of the most striking examples are on display in the Smithsonian Institution's National Museum of Natural History. In its magnificent Janet Annenberg Hooker Hall of Geology, Gems, and Minerals, more than fifteen hundred of the most spectacular specimens from the National Mineral Collection are displayed. The Hope Diamond in the Minerals and Gems Gallery is surrounded by other beautiful rock and mineral specimens, many of them from the Franklin and Sterling Hill mines. Visitors to the exhibit also will find many samples from Franklin and Sterling Hill in the Rock Gallery, where the role of water, heat, and pressure in creating the world's minerals is explained.

Many of the miners and other employees of the Zinc Company appreciate the diversity and uniqueness of the minerals that came from the Franklin and Sterling Hill mines, and some accumulated noteworthy collections. Over the years, many visitors came to Franklin and Sterling Hill to see, and often to collect, the minerals at the place of their origin. The great number of different minerals found in the ores, the world-famous reputation of the minerals, and the flourescent nature of some of the minerals has made, and continues to make, the Franklin-Ogdensburg area a special destination for rock enthusiasts.

Several of the narrators attempt to explain the charms of rock collecting in general and the particular delight in collecting Franklin and Sterling Hill minerals. For Steve Misiur, who now serves as curator, miner, and newsletter editor at the Sterling Hill Mining Museum, the enthusiasm started early. Steve remembers:

> [I had] an interest in mineral collecting ever since I was a young boy. Part of that interest is on my mother's side. Her father was a coal miner. He came from Poland, back in the late 1890s. Came from Poland to the United States and worked out in the coal mine in McKeesport, Pennsylvania. He was the only relative that actually worked in the coal mine. With that kind of background, a hard working Polish family, and being raised in Elizabeth — and with my natural curiosity of things around me — I developed an interest about the age of ten as a Pebble Pup, a little boy pecking away at the pebbles. Right in front of my house, by the sidewalk, is a little clay patch and there were pebbles in there. Then there was a little quartz crystal in the clay. It was very worn by glacial action, but I could see it. I was very curious. That was a virus so to speak and I bugged everybody that was around me, "What is this thing?" Nobody knew the answer. So I had to go to the public library and look up old books on rocks and minerals. And, sure enough, I flipped open one book and there is a drawing of a quartz crystal. I said, "Eureka!" I knew on my own what this was. I knew what I had. That propelled me into the interest of mineral collecting.
>
> At about the age of twelve, I got to collect on the Buckwheat Dump in Franklin. But that was not when everything was set up. This is 1966. They were just starting the Franklin Mineral Museum. When I started collecting on the dump, something was different than anything I had seen previously at that age. Somebody had a portable light and I got under a

canvas top and I was impressed with what I saw. But I was more impressed by the daylight texture of these rocks. I was just so interested.

But I stopped collecting about the age of fourteen, as I was busy being a kid, which was a full time activity in itself. I went to school. Graduated from high school in 1975. Then went to study architecture at a place called Mercer County College. Then I transferred to Pratt Institute in Brooklyn, where I got my Bachelor's in architecture in 1982.

Well, in 1983 and 1984, there was a recession; so therefore, I couldn't stay in the field on a full time basis. I basically wandered around within the field, trying to get a full time job. So, to keep myself out of trouble, I decided to try to get back into one of my old hobbies, mineral collecting. I came to this mine [Sterling Hill] to apply for a job, but a gentleman who was working there said, "It's very unlikely that you're going to get anything here because the company [Zinc Company] went from Gulf & Western to Horsehead Industry." Horsehead Industry did not have a farsighted mining plan here or policy for extending the life of what was here. "So," he said, "Come back in six months." I came back in August or September of 1983. He said, "No, we're not hiring anymore." He says, "Very likely within a few years, this whole operation will shut down."

Figure 47. Narrator Steve Misiur. "I developed an interest [in mineral collecting] about the age of ten." (Photo: Carrie Papa)

When I got back into the hobby. I rode up on Route 23 to get to the Franklin Mine. I went to the museum. Took the tour there. Went through the streets of Franklin. Got to see the old people. I rode home on Route 23, and there was as a rock shop in the town of Wayne called Jim's Gems. In that rock shop was a young lady by the name of Maureen Woods. She's working in our gift shop [at the Sterling Hill Mining Museum] today. Jim's Gems had a lot of minerals from around the world, but his specialty was Franklin-Sterling Hill minerals. I wandered around the store and I always kept going back to that case. Then Maureen says, "You like that stuff?" I said, "Well, it's different." I said, "It's a lot different than what I see from other localities and places." Then she started talking about the Franklin/Ogdensburg Mineralogical Society. That's the mineral society up here that specializes in the minerals of this district. She gave me an issue of *The Picking Table*. And she said, "Why don't you join?" It's only like, I believe at that time, seven dollars. She said, "You'll meet a lot of people." And she tossed out names like Dick Hauck, and Dick Bostwick, and Bernie Kozykowski. So we talked some more, and I said, "Well, I'll think of it. Thank you." And I left. And on my ride home, I'm thinking, "Well, why not? What do I have to lose? Just my time and seven dollars."

So I went to the next meeting they had, and I was hooked! I got into their club activities. I met Richard Hauck. Dick invited me to his house when he was living in Bloomfield. So Dick and I and his wife, Elna, became best friends. Then, when he and his brother bought this place, [the Sterling Hill Mine] back in 1989, he says, "Well, there's a place here for you anytime. Scrape paint, take out the garbage." At that time, I said, "Yeah, why not?" So I've been here ever since.

I was just here as a volunteer, doing whatever work they wanted me to do here. Just being here was a privilege itself. So Dick evidently saw that I was dedicated. One of the jobs that Dick gave me was to work with John Kolic to blast out a new tunnel. So, after the tunnel was done, I was talking with Dick, "Is anybody here doing any serious sampling of the rocks underground?" You always have to keep in mind the water was rising at an exorbitant nine inches a day. That meant that all of these places that are underground where you can see ore will be one day lost completely. So that became John's job and my job to be collectors of these minerals and geologic things in the mine. I became more or less the defacto curator of the geology and mineralogy. You can see by looking around these rooms here all these big slabs of rock. This is all of my four years of collecting these samples underground. We were racing against the water trying to get all that stuff up to the surface. What I'm doing here is a once-in-a-lifetime chance. There's some wonderful, wonderful things about this place, this history, the time that's past, the time yet to come. What you see here in my office and outside of the office, to the general public it has almost no value. To the average person in the street, what we have here is just nothing more than glorified pieces of doorstops. But, to the educated person, particularly to the scientist, or to the advanced mineral collector, they have value to that person. Franklin-Sterling Hill mineral collecting is very specialized. Let's say, I was to tell you that there was a franklinite crystal, let's say two inches across, which is very rare. For that size, I'd put a price on it you might be able to buy it. But the advanced mineral collector, you know, they would pay a huge amount of money. The advanced mineral collector, oh yes, they'd kill each other for it.

Let me give you a little anecdote. I was collecting in Franklin with a group of friends at a place called the Mill Site. It's the site of the former mill in Franklin. At the end of the day, we were talking about what minerals we had purchased from mineral dealers recently. And we were tossing amounts back and forth, "Well, I bought such and such for a buck and a half." And, "I bought that such and such for two bucks." And I said, "Well, I bought this one for a buck and a quarter." And there was a woman who was not at all a Franklin collector, but she was collecting that day. She says, "Excuse me, where do you get all of these wonderful minerals at such cheap prices?" We all looked at her, and she goes, "Well?" We said, "Excuse me, Madam, but when we say a buck, we don't mean one dollar, we mean one hundred dollars." There is a language within the mineral community that we use. When we say a buck, we mean one hundred dollars. And a thousand dollars is a big buck. So there are collectors in this hobby right now, in this Franklin-Sterling Hill field that will spend anywhere from five to ten big bucks.

When you think about it, my hobby is my job. I'm here doing what I like. Nothing that I ever will design as an architect will last as long as that tunnel. It will be there ages after I'm gone. There will be people still walking through there. They will still be getting a hint of the wonder that John [Kolic] and I got of being the very first men to see this rock no human eye had ever seen before. Yes, there is a sense of accomplishment. Pride.

Former geologist and present curator of the Franklin Mineral Museum, John Baum, discusses mineral shows and prices. Although John no longer accumulates specimens for his private collection, he is still involved in purchasing and labeling rocks for the museum. John declares that his interest in rocks and collecting minerals goes back to early childhood.

I started collecting when I was ten years old, if not earlier. I used to spend my allowance buying minerals. I got quite an attachment to minerals. First of all, I was interested in the out of doors. And, secondly, my grandfather had been interested in mining, although he passed away when I was four years old. Up in the attic, he had a lot of ore specimens that he picked up in Colorado and they sort of intrigued me. My older brother took a liking to those and he and I took the collection. Brought it back to my mother's house where I set them up in a dresser drawer, which was the beginning of my collection.

One drawer for minerals, one drawer for seashells. I didn't particularly favor one over the other at that time, although the concentration on minerals came very quickly, as soon as I started buying them with my allowance. Unfortunately, I bought as much as I could with as little money as I had, so I ended up with the fifteen cent pieces and the twenty-five cent pieces, and today they're not worth a great deal. But they are worth three or four times what I paid for them. What I should have done, of course adult thinking, is to have saved the money and then buy one really good piece, but what does a child know about it, really. So I had a collection from early on and it had a few Franklin minerals in it that I bought from dealers. Once I got here, of course, I picked up specimens as I went along, so I had a pretty good collection. I've turned that over to the Smithsonian now. They have that collection. But I still have what we call here foreign minerals, minerals from outside the Franklin area. Those I am slowly bringing to the museum for them to sell for their own benefit.

We have some stuff for everybody. These are one and two dollar specimens. [Looking at a box of specimens], here's a fifty-cent tag for goodness sakes. Hardly worth my time to write a tag for fifty cents, but somebody will buy that. Every one that we sell, I've got this little tag for the treasurer too. That comes off of that. We have to keep track of whose collection it came from. This is a rhodonite. Women like pink stones.

I'm trying to think what is the most I've heard offered for a Franklin specimen. I'd say fifteen thousand dollars. I've got one fluorescent specimen that's supposed to be worth over five thousand dollars. The rarity of it, you might say. All really good artistic specimens are way up now. Forty thousand dollars isn't too much to pay for a chunk of rock. If you got good crystals out of it. That's beautiful stuff.

They are mining in Colorado just to recover specimens for that particular reason. There are people that'll mine for a year and then take a small number of specimens, really good stuff, to the Tucson Mineral Show and sell them. People come from Germany to sell stuff there or just visit. And the great collectors that come from Italy and Spain and so forth are there. Big business. The Tucson show itself goes for four or five days. The Franklin Mineral Show is fairly well attended. And the Franklin show is [merchandising] when you get right down to it. Let's face it. There are cases of minerals at the Franklin show to see and people do spend time looking at them, but the big thing is the opportunity to purchase.

Former Franklin miner Nick Zipco has one of the most comprehensive private collections of Franklin and Sterling Hill minerals in existence today. In appreciation for his work in acquiring and preserving the Buckwheat Dump, for managing the Trotter Dump for many years, and for sharing his knowledge and enthusiasm for the local minerals, Nick was honored by having the 1995 bulletin of the Franklin-Sterling Hill Gem and Mineral Show dedicated to him.

As with Steve Misiur and John Baum, Nick's interest in collecting started at an early age with a gift of two specimens from the Franklin Mine. Nick was living in Pennsylvania where his father worked in the coal mines and his mother ran a boarding house. One of the boarders had moved to Franklin and, as Nick says:

He came back to Pennsylvania for a visit and he gave me these two rocks. I held them in my hand up in the air, and I'm looking at them and I said, "Rocks? What are rocks?" This is [when I was] seven years old. I found out that they were two rare rocks, so that started me to look for different kinds of rocks in this area. When we came here [to Franklin] I was twelve. Then, when I got in the mine, I had opportunities to go to any level in the mine. I did all kinds of mining. I started in 1936. What I did on my lunch hour, I ate it as quick as I could and I went looking for minerals, different places where I hear, "Something's coming out of here," "something's coming out of there." I'd try to get in them places and get certain rocks.

There was no objection. My boss used to carry it up. I had a good boss. He carried it up dinnertime. I couldn't go up dinnertime, but they went up in the change house. He'd put it in my locker. Then I'd give him pieces. To recognize one mineral from another, I learned by hand. I've got minerals here nobody's got.

I nearly got killed twice looking for rocks down the mine. I never got hurt working, but looking for rocks. The one time, I go in the place and they had cleaned it out. They fired and mucked everything out. I had a spotlight. Battery light. Far away, I seen a certain rock over there. So I'm going in. I didn't look up. I'm going right in there and pieces are hitting me on the head. I look up. "Oh!" I wasn't back a minute and this big timber, eighteen foot long, come down! It was black. For fifteen minutes, I couldn't see nothing in there, it was so black. The dust. See, what they did, they went on lunch. They cleaned this out ready to put in a setup. And it wasn't there. I walked in before that setup was up. They went to get dinner. I got the rock! It's worth about three hundred dollars.

I've got a rock in here that cost me five thousand dollars! Took me forty years to get from the guy. From the miner. It's valuable because this stuff here, this white stuff, this is the rarest stuff in the mine. Roeblingite with ganophylite. This is a rare mineral on top of this brown. I have people come here and they say, "Why don't you sell me them rocks out here? Don't you want money?" Nope. I don't want to sell them. I could be a rich man if I wanted to sell them. I don't want to. Money can't buy my rocks.

Figure 48. Mrs. Zipco and narrator Nick Zipco. Former Franklin miner Nick Zipco has one of the most comprehensive private collections of Franklin and Sterling Hill minerals in existence today. (Photo: Carrie Papa)

Dick Hauck's interest in minerals was cultivated by the Bloomfield school system when the principal arbitrarily assigned him to be a member of the school's mineral club.

> "You, you, you, up to Mrs. Sherlock's mineral club." That was the start of it. I was shanghaied. Got exposed to a fascinating pursuit. The treasure hunt of every kid's dreams, going out there, roaming the hills, looking for these hidden treasures. We never outgrow it totally. So that particular start got me interested in minerals. I shared it with my brother — who is eleven years younger than I — as he started to get to a stage where he could go out and explore various areas.
>
> My interest developed further. Eventually, here I am working in the area that so many people worked hard to preserve. The Franklin Kiwanians got a show going in Franklin many years ago. Then two mineral societies were started. I was influential in the second mineral society — the Franklin/Ogdensburg Mineralogical Society — and, because of circumstance, was actually their first president. My qualifications were: number one, I was there, number two, nobody else wanted it. [laughter] The society got born and has survived ever since through some interesting and challenging times. Then the society influence associated with the museum influence. Now we have a very successful show

in the area. Now we have two museums in the area. We are seeing a very strong and healthy attention to the minerals and mining history of the Franklin-Ogdensburg area. Our mineralogy is unique. The fluorescents being the world's best. Our rocks are beautiful.

Architect and former miner, Bernie Kozykowski, is yet another person whose interest in mineral collecting developed while he was still a child. Although Bernie was not a Franklin resident, his uncle, Evan Jones, worked in the Franklin Mine. It was through visiting his Aunt Flo and Uncle Eekie that Bernie came to know Franklin and its minerals.

I remember coming here as a boy when he [Uncle Eekie] was still working in the mines. The mine shut down in 1954. In fact, he lived in a company house right up here on Main Street. I remember playing in the yard with my cousins. We would tour the town and see different things. We were pretty young at that point. I was nine years old. My first recollection of the minerals in Franklin was the fluorescent aspect of them. I think [my interest] was more out of curiosity than anything else. As far back as I can remember, I've always been curious about most things, particularly those things that occur in nature. I mentioned early on that my first exposure to the minerals here in Franklin was the fluorescent aspect of it. Yet, at the same time, I wasn't captured by them at that particular moment. In fact, I remember my uncle, who was working in the mines at the time, giving me mineral specimens to have my own small collection.

Being naive and unenlightened at the time, I gave them to friends and they're gone. He's criticized me to this very day for having done that. But you're talking about a child in grade school, grammar school, and there wasn't a whole lot of understanding of science. As I headed toward junior high school, I started becoming aware of things in more technical terms as to their place in nature. Then I was reading a book in a library — I used to hide out in the library. I grew up in a tough neighborhood and the library was my escape chamber, so to speak. I happened to be leafing through this one small book on minerals and I came across Franklin and I said, "Wait a minute, my uncle lives in Franklin. Maybe there's something to this." At the same time, the Kiwanis Club here in Franklin started the mineral show that occurs here every year. This was back in the mid 1950s. My uncle said, "There's a mineral show in October. Why don't you come on over?" I did and he took me to the mineral show. I was probably eleven or twelve years old at the time. I saw the exhibits. One thing led to another, and I came back and started collecting minerals and learning about them.

The fascinating thing about the mineralogy here is there's so much of it that you can still find some very interesting minerals. In fact, within the past five years, when they were digging around on the collecting area next to the Franklin Mineral Museum, they actually came up with a brand new mineral species there, new to science. So these things occur. It's not over. It's a very exciting thing yet.

The two places to collect here originally were the Parker Dump and the Buckwheat Dump. The Parker Dump was located pretty much where the current firehouse is. At the junction of High Street and Buckwheat Road. That was the location of the works of the Parker Shaft, which was one of the first shafts that went underground in Franklin. That was put down in the late 1880s or early 1890s. When they excavated for the shaft, they threw the waste off around there. And, as they went into the workings themselves, they would bring out the ore. They actually separated the ore at what was then known as Well Number One in Franklin, at the Parker Shaft. At that head site. They had a mill there for processing the ore, strictly for separating it. At this point, it was shipped out for smelting to the north, to Bethlehem or to Palmerton, Pennsylvania. At the same time, the waste rock that didn't have value was just discarded. Huge piles of material accumulated. So you have a dump, literally waste rock from the mine. The same thing occurred along Evans Street, at the Buckwheat Open Cut. When they were mining there, they took the waste rock and simply dumped it over the side. These were the two principle collecting areas.

What makes these minerals so special? It's the uniqueness of it, everything about Franklin. There isn't another place like it in the planet.

Geologist Bob Svecz's father and two grandfathers had all worked in the Franklin Mine. Bob was only two years old at the time the Franklin Mine closed, but he had heard about the mineral wonders of Franklin all of his life.

There was always a kind of fascination about it because everybody talked about the mine, and of course, the minerals. Neighbors. Parents. Friends. Relatives. My grandfather had some minerals, but he really, in a sense, wouldn't know what he was looking at. He wouldn't know an allactite from a sarkinite, a hardystonite, esperite, fluorescent minerals like that or a piece of native copper or a vein of rhodonite, which had a bright pink color to it. He'd pick up things that were pretty.

What made me go into geology? To be honest about it, it was time I had to go to college and I had never thought about it. But I had to pick something. So I just figured, rocks, minerals, okay, I'll choose geology. It is fascinating. I did find it to be fascinating. From 1971 until 1975, I worked as a miner's helper [at Sterling Hill] during summer months when I was off from school. Then [after graduation] I didn't have any money and this was one of the better paying jobs around in the area. So instead of going out looking for a job — let's say I would get called for an interview out in Akron, Ohio. You needed money to travel out there. I didn't want to burden my parents anymore, so I figured, I'll spend some time working, getting some money, seeing what it's like to actually go out and work for a living.

While I was doing that, the company saw the need for an active geologist at this mine. The biggest reason that I could see for that is because we were nearing the end of the life of the mine. All the "gravy" ore, all the ore that was easily mineable, had been taken. It's coming down to the end. We're scraping the bottom of the barrel. We were exploring something on one of the lower levels in the mine when the mine closed. It could have turned out to be something really good or it could just have been an isolated pocket.

Working as a geologist, I could get to other places in the mine. Working as a miner, you were only supposed to be in the places where you were assigned to work. I know of a lot of other guys, when they had the time, they would wander around other places and look for minerals. That's why mineral collecting was frowned upon.

It got to the point later on where guys were carrying large chunks on the cage out of the mine. They [the mine officials] were concerned that if some got dropped from the cage when it was moving, it could be dangerous. So it was issued in writing that the miners were allowed to collect mineral samples, but only those that would fit in their lunch bucket. That's the only time that it has come out, as far as I know, written permission for anybody to collect minerals. That made it kind of legal. Before that, there wasn't anything said about it. For myself, as a geologist, if I can carry it or roll it from the station on to the cage and bring it up, I could take it. But most of that, it stayed there.

According to Tom Sliker, former mine superintendent, many people worked as miners at Sterling Hill for the express purpose of collecting minerals.

It got so there was a lot of people came here just to get a job to get minerals. You'd be surprised how many people that has no use for mining but they love to be here to get the minerals. At one time, they tried to stop it, but the people would take it out in their lunch pails, you know. To me, it's just a rock. I got so used to seeing the ore. But people like the light, the fluorescence. You have a green light. You have a red light. You have orange colors. You have purple colors. There's all colors and I think that's what people like. There's no other mine except Ogdensburg and the Franklin mine that has minerals that fluoresce like ours does. They've become quite expensive because there's no other minerals in the world like Franklin and Ogdensburg.

Miner John Kolic readily admits that he took a job at Sterling Hill because of his desire to collect minerals.

My interest in minerals and collecting on the dumps here [in Franklin] led me to take a job in the mine in Ogdensburg. Now my primary interest in taking a job in the mine was to find minerals for my collection. Minerals that I had read about, willamite, franklinite, interesting things. About 1969, I was living in Arizona and I took a trip to the Grand Canyon and hiked down the Bright Angel Trail to the bottom and back out. As you walk up and down the Bright Angel Trail, you get a view of the number of different types of rock and I got interested. So I bought a used college textbook about geology in order to learn something about what I had seen. Subsequently, I came back here to New Jersey. At that time, they were doing the rock cuts for Route 15 and Route 80. I went out there to see if I could recognize in the rock what I was reading about in the book. I had a general vague interest in natural history, but nothing specific. Then, when I went out to those road cuts and looked at the rock types in the walls of the cuts, I picked up samples of rock. And I said to myself, "Gee, what are these minerals in here?"

So now I'm interested in minerals and I had to buy a book on minerals because that covers minerals more thoroughly than the geology book. I picked up a few rocks and I took them home. One day, I was unloading the rocks out of the back of my truck and a guy next door sees me and he says, "Oh, you're interested in rocks, huh?" He tells me, "A friend of mine has a fireplace made out of rocks that glow in the dark. He gets them from Franklin, New Jersey." I says, "Where's that?" He told me how to get there. I came up here and started hunting on the dumps and just got hooked.

I got a good collection. They [the Zinc Company] didn't encourage it, but as long as you did your job . . . . After about the first year, some of the bosses became aware that I was taking out a lot of rocks so they checked my work record, my performance, how I was producing. They decided that I was doing my job so they weren't going to hassle me about the rocks. Later on, about ten years later, they actually came out with a standard. They published a notice on the bulletin board that they didn't mind if miners took out samples of rock from their own working areas for their own use, their own collections. But the piece had to be small enough to fit in a lunch pail, and they didn't want you wandering to other levels where you may have heard something good was being found. They didn't want you wandering off somewhere else while you were supposed to be working at your place. As long as you stayed within those categories, they tolerated it because it was a matter of morale.

After many years of experience, you learn to recognize by eye things that you had no clue as to what they were in the beginning. Some of the things, once you've seen them once, you have no trouble recognizing them again. Others — for instance there's maybe twenty-five or thirty black minerals here — it's hard to distinguish between them. Some of that stuff you actually need a scientific examination to determine what it is. As an experienced collector, as a miner who did the work of breaking the rock, you know whether you're working in ore or rock. The ore is pretty easy to tell. First of all, it's twice as heavy. If you pick it up, you notice it right away. But there were times when you would go through a drought where there's maybe six months where you never came across a good specimen. Sometimes when you're going through a period like that, you may take home chunks of rock just for the sake of taking something home.

A lot of miners, that's all they did was carry rock home because they heard that rocks were good. So they carried some home. They had no idea what they were and they didn't really care. They just had to find a buyer someday. And it turns out that they had nothing, so they didn't find a buyer.

Ewald Gerstmann is one local person who did a lot of buying from the miners. Ewald amassed such a grand collection, he had to build a museum in which to house it. Eventually, the Gerstmann collection was donated to the Franklin Mineral Museum, where it remains today. In the following, Ewald describes how and when his interest in mineral collecting began.

I wasn't interested in minerals till the New Jersey Zinc Company closed. If you remember back in them days, the miners, they had minerals, but they really didn't know what they had. They just accumulated them. So I started buying from the miners. At the time when I was buying, I bought a lot of material because in them days it was cheap. You have to remember that the miners only made, every two weeks, about twenty-six dollars. Once they know that they had a sucker to buy it, they came here. Once one of the miners sold me something, a pile of rocks. They thought we were crazy. Two other miners had come the next day because they talked about it. So we just bought.

Then when I had a large accumulation of rocks and minerals, I had a scientist come in here. He wanted to buy some of them. It was a doctor that come. Doctor of Mineralogy. Dr. Frederick Pough. He came and he wanted to buy some. For a piece that I paid five dollars, he offered me a hundred. So that's more than I made in them days in a month. So I figured if it's worth a hundred to him, it's gotta be worth a hundred to me. In a case like that it encouraged you to collect more and buy more. That's better than working for a living.

So when these types of doctors come around, and then they go back in their circles, they must have talked. He used to be the curator of the Museum of Natural History in New York. Right after that, I had quite a few more come here. Then more came, they started identifying the stuff for me, not just by sight but by the chemistry. I'd give them samples so I'd find out what I had. From there, I started on the collection. The real collection. And, when I had a lot of accumulation, I sold it to other collectors. Just about every miner brought home stones. They didn't all bring good stuff. A lot of them just had rocks. I know more than the miners. You pick up. And with the scientist, you pick up.

I was doing this for about six years. I got out of the plumbing and heating business and I got this [building] to make my own museum. Prior to this, prior to me being here, it was a carpenter shop. Nothing in here, just a blank building. I bought the property here just to be a museum.

I didn't sell them at that time. I was just buying them. The number of different miners that I bought from was three to five hundred. After you get the knowledge, you get very selective. You only maybe buy one piece. When you get involved in something, it becomes an obsession.

Figure 49. Narrator Ewald Gerstmann. "The number of different miners that I bought from was three to five hundred." (Photo: Carrie Papa)

I'm not involved in collecting now. When I sold the collection, I didn't make any money on it. I lost money. If I would have had that today, I could have lived the rest of my life like a king on what I could get. The most satisfying part of having this collection was the accomplishment, which I never thought of when I started. To have amassed this many in one spot, I have to give credit to the miners. All the miners.

Assistant manager of the Franklin Mineral Museum, Steve Sanford, has written several articles for *The Picking Table*, the journal of the Franklin/Ogdensburg Mineralogical Society. Steve worked for the Zinc Company during the 1970s and collects minerals from both the Franklin and Sterling Hill Mines. When Steve sold his collection, it had been appraised for around thirty-eight thousand dollars. Since that time, he's been told its value has more than doubled. Today, Steve maintains a collection of rocks that describe the ore bodies in hand specimens, or as Steve says, rocks that tell a story. In his interview Steve explains:

> Mines fascinated me. I was always kind of interested in the history of the earth. Seven years prior to working there [Sterling Hill], I'd become fascinated with Franklin mineralogy. I originally came up here looking for a mine to explore. I'd been crawling around the iron mines in Morris County, underground. I was looking for a mine crawl up here. But it [the Franklin Mine] was full of water. I stopped at the mineral museum in 1968, and got a copy of Palache [Charles Palache — crystallographer, professor, author]. I got hooked reading it, and I've been deeply interested ever since. I worked for the New Jersey Zinc Company in 1970 until July 1975. Then I went back to school. I was at the Franklin Mineral Museum when Sterling closed in 1986.
>
> But I did enjoy working at the mine. I liked the upper section. It was cooler. Better ventilation. Some kind of ores — it sounds a little silly — but some kinds of ores were depressing to me to work in. I didn't like working in a black ore stope. I was a mineral collector and black ore didn't have that much. Grunge looking. The other ore could be stunningly beautiful. Sometimes rocks look better underground in mine light than they did on the surface later when you brought them out. The Sterling Hill ore body had a complex physical

configuration, and a complex chemical makeup. One part of the ore body was characterized by having black willemite and black franklinite.

[Pointing] That case there has rocks that have a story to tell if you can interpret it. All of this is ore. There's no waste in this, almost. That's what we used to mine. It's zinc, iron, maganese ore. Its primary worth is its zinc content. Iron, maganese also. For most of the history of the mines, the iron maganese paid for everything. Paid all the expenses, mining, milling, transport, everything. And the zinc was all gravy. That's why they made such incredible windfall profits there.

Now, these two rocks are the exact same thing, willemite and franklinite. Neither one has much waste in it. This is what I call black ore. This is a different piece of Franklin ore. The main difference is the willemite. The willemite's pale green in Franklin ore. At Sterling Hill, it's mostly altered. This original light colored willemite was partially broken down so it looks like red in most places. In that one place it looks real black. These originally may have been similar in appearance. This ore got this appearance and it stayed the same appearance basically for eleven hundred million years. That's twice the period of organized life on earth. At some later time, they were altered to look like red and black. I could spend hours with those rocks describing the stories they have to tell. Processes. I have faults. I have collapsed stretches. I have reactionaries. Each one of them has a story to tell. I'm trying to save some of this before it is lost to science forever. These are mostly process specimens. Processes that happened.

A Franklin collector is, I feel, a cut or two above a rock hound because many of the Franklin rocks, which are highly prized, are not aesthetically pleasing. They are more a pleasure of the mind in knowing that something so complex is so rare or so unusual. You have to be able to appreciate some of the inner value of something rather than the external. A cut diamond is very valuable, but it's more a pleasure of the spirit. These things are a pleasure to the intellect.

Now that the Franklin and Ogdensburg mines are flooded, we have lost the opportunity to obtain additional minerals from their original site thousands of feet below the surface of the earth. When Sterling Hill, the last operating mine in New Jersey, closed in 1986, it was the end of an era. Anyone connected with Franklin or Ogdensburg would agree with Bernie Kozykowski when he says there was "something very special here."

Part Two

The Model Mining Town of America

Franklin Borough is primarily the creation and outgrowth of the mining industry as developed at Mine Hill in the last thirty years. A public park, a public swimming pool, an athletic field, good roads, street lights, and water supply, are all factors that make Franklin Borough a thriving, contented, and happy community that is sometimes called the model mining town of America.

Historical Notes of the Iron and Zinc Mining Industry
Sussex County, New Jersey

Big Brother or Patron Saint

In 1913 Franklin was a company town, sharing the benefits and the drawbacks of the Zinc Company's paternalism.

Franklin Borough Golden Jubilee

Many historical studies portray the company-owned mining towns of America in a negative light. According to these accounts, the communities were dominated by ruthless mine owners who exploited their employees and provided them with appalling living conditions. Company housing often consisted of row upon row of unsightly standardized buildings constructed of unpainted lumber, cement blocks, or other inexpensive materials. In 1922, when the United States Coal Commission investigated the standard of living of miners, it found that of the 664,000 coal miners employed across the country, more than half were living in company housing; of those, only 14 percent had running water and only 3 percent had a bath tub or flush toilets. [1] Twenty-four years later, another government report, *A Medical Survey of the Bituminous Coal Industry*, found little improvement: fewer than half of the houses in the survey had piped water and 75 percent of the company towns still used outhouses.[2]

The community facilities of company towns also were criticized as being inadequate or inaccessible. Schools, fire departments, hospitals, and stores were limited and few recreational amenities, such as theaters, parks, and libraries, existed.

The prevailing images in historical accounts are of a controlling mine owner and impoverished residents. It has been argued that a one-industry town allowed the mining companies

147

to exert total economic and political control over the residents. This imbalance of power, in some cases, led to unrest and violence among laborers and the companies' attempts to thwart unrest, uprisings, and unionization.

Contrary to the negative condition that existed in some mining communities, there were corporations that were concerned with the well-being of employees and residents. The Phelps Dodge Corporation in Arizona, Kennecott Copper in Utah, Calumet & Hecla in Michigan, and the Stonega Coke and Coal Company of Virginia provided quality housing and facilities; these companies operated with the belief that it was in their best interest to keep the resident miners content. These operations and several others were often viewed by sociologists as "ideal communities." Butte, Montana was billed as the "World's Greatest Mining Town." In New Mexico, Phelps Dodge hired a New York architect to design its "model community" of Tyrone, which a 1917 report by the United States Department of Labor concluded was "exceptionally well maintained."[3] In Michigan, Calumet & Hecla was turning its mining towns into the "most go-ahead places on the Upper Peninsula," while company management was praised for "providing almost everything workers could desire."[4]

Figure 50. Interior of a miner's home in Bell County, Kentucky, 1949. (Photo: Courtesy of the Bureau of Mines Collection, National Mine Health & Safety Academy)

To understand the effect that company-owned towns had on individuals, historians conducted a number of interviews and surveys, which were compiled in *Appalachian Coal Mining Memories* and were also made a part of federal government reports. Significantly, the sentiments recorded in those documents are echoed by the narrators who took part in the Mining Oral History Project. The miners employed by the New Jersey Zinc Company who lived in Franklin and Ogdensburg, therefore, provide a representative view of how miners saw themselves, their community, and their employer.

In the following, Steve Revay remembers Robert Catlin, a prominent mining engineer who is credited with modernizing the Zinc Company and improving living conditions. Steve also highlights some of the more domineering qualities of the company.

> You know, when they started to mine, they didn't know how to get this ore the hell out of here. They knew there was ore in this mine, but there was four or five guys, I forget how many, had rights in this whole thing. And they didn't know what the hell to do. They couldn't get together. But finally, they had one group, the company [the New Jersey Zinc Company]. This group knew they had the mine. They knew they had the ore, and then they knew they had the legal right to go ahead and mine it. But they couldn't find anybody that knew enough about it to start a mine. So they scouted around and scouted around. Finally, they found this guy Catlin, who was a very well known engineer in Africa. He was a great friend of Herbert Hoover, the President. They were buddies, Hoover and this Catlin. So they brought this Catlin guy to Franklin and he's the guy that showed them how to get this ore out of the mine. He's the guy that had the shafts sunk and had everything fixed up so they could go ahead and mine. He had a system to mine the ore to get it out.
>
> He came from Africa. He lived in that big home in Franklin below where the nursing home is. They built that for Catlin. Catlin had a chauffeur. He had everything. Hell, they took care of him, you know. And he made damn good money. But, he's the guy. If it wasn't for him, they wouldn't get the ore. Hell, they had holes dug all over the place, and nobody knew how to mine it.

When the Zinc Company took over the mine, Catlin had them put their own water system in. They built a filter down there and they started to put water lines in all through the town. They built homes. They wanted to get that ore out, they had to have workers. They built homes for them. Catlin did that. And you didn't have to pay for your water. You got your water free. The Zinc Company also made their own electricity. In the beginning, we had kerosene lamps. Water, you had to go out to get. They had a pump here and there. Then Catlin supplied them with water. They supplied the whole town with electricity. They put electric lights in town. And every once in awhile, you'd have a free month when you didn't have to pay for electricity. You didn't have to pay for your water. When Catlin run the place, Catlin was a real gentleman. He had

Figure 51. Company cars and chauffeurs in the 1950s. (Photo: *Zinc Magazine;* Courtesy of Ogdensburg Historical Society)

compassion for everybody. He was a very nice man.

But when McCann took over, the first thing he did, you had to pay for your water.[5] So much a month for your water. You got no more free electrical months from the powerhouse. That McCann he was a shrewd cookie. He made a lot of moves to save money for the company. When Catlin left and he took over, the first thing he did, I remember, he had every employee that was working for the Zinc Company examined up in the hospital. They had to go through a physical. And all the guys that he found that had something wrong with them that would impair their work, he laid them off. That's how tough he was. Just laid them off without boo, you know. There was two or three guys committed suicide because they got laid off. They didn't have no work and didn't know what to do. Maybe the Zinc Company was all right, but the people running it was no damn good.

At one time they had people coming in here that wanted to organize the union. They never got organized because they run them the hell out. Then they didn't bother for a long, long time until they passed a law where it was legal to have unions. Well, then they couldn't run them out. Before that, they passed a law — you didn't have no union yet — but the law said, the men could go to the people that were running things and air their grievances, even though you didn't have no union. This is the law. So Franklin had a committee like that and they went to see McCann. He was the superintendent. Well, one of the grievances was in the change house. They had a lot of cockroaches, and they wanted the Zinc Company to get an exterminator in there to get rid of the cockroaches. Another thing, they run out of hot water quite often. When they were taking showers, they didn't have no hot water. There was quite a lot of other stuff that went on that they aired with their grievances. And you know what McCann said to them? He says, "The next thing you're going to ask for is automatic a-- wipers." [laughter]

There were other manufacturing companies that wanted to get into Franklin, but the Zinc Company kept them out because they would have competition in their labor. They didn't want nobody in here. I remember one candy company wanted to come in here, and the Zinc Company wouldn't let them in. Hell, they made an awful lot of money in this town.

It was a mining town. It was a company-owned town and the company run it. See, the borough and the Zinc Company were together because most of the councilmen worked for the Zinc Company. So, if the Zinc Company wanted something done, they just let a councilman know, "We want this, we want that," and they'd do it. The Zinc Company owned the whole town. They owned all the land. They owned all the buildings.

Did Steve [Novak] ever tell you about when he was tax assessor? Well, he run for assessor and he got elected. He began to go around and he assessed the properties. Then he found out that all the Zinc Company properties were taxed a hell of a lot less than the average citizen would pay. So he began to jack their assessments up. Munson was a great friend of the Zinc Company. His assessment was hardly nothing, and he had one of the biggest homes in Franklin. So Steve started to change the assessments on all of them. They [the Zinc Company] did every damn thing to get rid of him. To get something on him that would ruin his reputation and everything else. They tried to bribe him. They accused him of stealing money. They offered him a hell of a good job. They hired a lawyer to watch him. They tried to bribe him just to keep assessments down, but he wouldn't do it. He got elected by the people, so they couldn't get rid of him. It was funny.

At one time, the only way you could get in this town or get out of town was by train. I remember when there wasn't a car in town. If you wanted to get anywhere, you had to go by horse and wagon or go by train. The Zinc Company had four or five spies. They'd meet every train that would come in and they'd meet every train that went out, so they knew who went out of town. They knew who come in town. If there was any undesirables coming in town, they'd run them the hell out. And, if there was anybody in town trying to leave, you know, by not paying the rent, or owing money, they'd know that. They wouldn't let them out of town. They'd put them in jail.

But some of the things the Zinc Company did were good. Like the Neighborhood House. They did something worthwhile there because the kids had a place to go spend their time. They had dances there and all that. They had three women working there taking care of it. They had a library. They had a poolroom. They had four bowling alleys. They had the nurse who lived there. Took care of the babies. Miss Pancoast, she was the head of the Neighborhood House. The band used to rehearse there.

The Zinc Company, well, there's a lot of faults, but the one big fault was they were very, very low paying people. They never paid no wages. I was only fifteen when I started to work in 1922. Thirty-one cents an hour. Then, even as I went along, I never got above fifty-three cents. That's the highest I got, after thirty-one years. I worked for them thirty-one years, and I never got no pension. The Zinc Company was very frugal.

One time they were getting twenty-five cents a pound for zinc. I worked loading cars at that time. Sometimes, we loaded

Figure 52. The Zinc Company generated its own power and sold it to the citizens, giving them two months of free electricity each year. (Photo: *Zinc Magazine;* Courtesy of Ogdensburg Historical Society)

fifty, sixty cars a day and each car had forty ton in it, forty-five ton. Imagine getting twenty-five cents a pound for the stuff. And I was getting fifty-three cents an hour. After thirty-one years. That was the big detriment. There was no way to get ahead. And it wasn't because they didn't have money, because they were filthy with money.

The president of the Zinc Company, his name was Palmer. Palmer donated the Palmer Stadium, football stadium in Princeton. He donated that. That costs millions. I'd say anywhere between 70 or 80 million dollars, because you can't build a big stadium for nothing. He donated that. But most of the wealth the Zinc Company had come from Franklin. Now see, all that money donated to Princeton. They never donated nothing here, never left one cent for Franklin where the wealth all came from. All that was left here was a hole in the ground.

While all of the narrators are in agreement that the Zinc

Company was notoriously low paying, many feel the benefits of working for the company and living in Franklin more than compensated for the low wages. Former geologist John Baum reports on the influence of Robert Catlin and observes that the New Jersey Zinc Company served as a model for other mining communities.

Although John appreciates the services the Zinc Company provided for their workers, he acknowledges that the influence the Zinc Company had in the town resulted in a controlled environment that would not be tolerated today.

> This was one of the best [mining towns] in the country. This was a prize. This was the place to be. Yes. It was a model mining community. Definitely. Some of the young men, of course, were disgruntled because the pay was not very high. But it had so many offsetting factors, which many of them appreciated, but some of the young men did not appreciate. They knew that elsewhere you could get more money. I knew I could get more money, but I could never get the living conditions I had here. The Zinc Company was extremely enlightened in regard to its treatment of the employees. It was of course profit minded, but it was very paternal — Franklin being the best example of that.
>
> The mines were consolidated in 1897, I believe it was, in what then became the Zinc Company. They realized they had a big problem, a lot of little mines and they had no system of mining. It was a mess. It was very difficult to do any more mining in Franklin near the surface because the place was full of tunnels and if you took out any more ore, the whole thing would cave in. So they sought the best mining man that they could find to solve the problem and they ended up with a fellow by the name of Robert Catlin. They built a fancy home for Catlin, which was later the boarding house where I lived on Main Street, with a garage, and living quarters for his staff, and a pool hall, and all the rest of it. They brought him over from South Africa. Catlin hired the best men he could get. That's when he sent to Michigan for Cousin Jacks. The mine boss, was a fellow by the name of Rowe, and they set up a new mining method.
>
> Then they established the town. This was following the consolidation, about 1906 and on. That's when this town was rebuilt. The company set up the town and saw to it that there was a bank here, a general store, a hospital. Catlin

observed that the miners that were injured lay on the platform until the next train for Paterson came along when they could go to a hospital. He said, "No more of that." He built the hospital. The first one in Sussex County.

Of course the streets were a mess. He paved the streets. He put in our water system. Certain areas subsequently had sewage. They built the homes and rented them out cheaply. My house, for instance, was a three-bedroom house and I rented it for twenty-seven dollars and fifty cents a month. It was a good company to work for in that regard.

Robert Catlin saw to it that all the homes had electricity. It was generated right down here at the plant. For years, I could not have an electric clock in my house because the current was made by the Zinc Company, and the operators were not too particular as to how many cycles were coming through this thing. Your clock would go fast one day and slow the next day. However, because the Zinc Company made its own power and sold it to the citizens, two of the twelve months of the year were free. The idea was to find out which month was free. If you lived in the north end of town and they started reading the meters in the south end and said, "Oh, incidentally, there's no charge for this month," we would get on our telephone and call up the other end and say, "This is free month, free month." And before the guy got there to read the meter, people would start ironing like mad and use up a little current.

The first year I was here, I was paid by the hour. That's when I was getting twenty-seven dollars and fifty cents a week. Fifty-five dollars every two weeks. From that pay, Social Security was deducted and our health benefits, things of that sort. After I was here one year and having proved myself, I was put on salary at something like a hundred and sixteen dollars a month, I believe it was, which was a raise. The raises were like that. Five or ten dollars a month and they came every two years. If you behaved yourself, you could expect a salary raise every two years, five, possibly ten, dollars a month.

There were times when I was so fed up with the Zinc Company, I was close to a nervous breakdown. The Zinc Company would not hire anybody over forty years of age, because after that you had twenty-five years to work to justify a pension. So the day you became forty, you were frozen into your job, unless you wanted to go out and start all over again with a job in a company you knew nothing about. The day I got to be forty was the worst day of my life. I had worked sixteen years at that time. Something like that. It was frustrating because they didn't pay much.

But this was one of the first companies to pay pensions. We had pensions long before Social Security was thought of.[6] But, because they guarded their money very carefully, when Social Security came in, they deducted that amount from the company pension. So many men were getting pensions I have heard of as little as five dollars a month. When I retired, I received a very modest pension. If you retired at fifty-five, your pension was discounted 6 percent a year. You were supposed to retire at sixty-five, so I lost 60 percent of my pension. Now, I had thirty-two years mind you. Any other company, you had thirty-two years today, you're not going to lose any of your pension. But in those days things were different.

The whole town was built by the company. I don't want to make a big deal out of this, but the Presbyterian Church was run by the salaried employees. And the Franklin School was run by the salaried employees, because they were all members of the Presbyterian Church. The school, of course, was established by the company. The mayor of Franklin, for, my God, it must have been over twenty years, was a chemist for the Zinc Company. Again showing you the hold that the Zinc Company had on folks in the area. They did court control of the elections. Same group controlled the church and the school, as I say.

The Zinc Company was very scared of unions. When I came here, we had no union, but the company had a company detective. It was his job to see who was a stranger in town. The company was so serious about this that they even took the numbers off of low flying airplanes to check as to who was making a survey of the town and for what purpose. The company detective was dearly hated by many, many people.

Still, I would say it was a VERY good life living in Franklin. Womb to tomb, people were pretty well taken care of because this was the most modern town in the county. Everybody knew they didn't pay much. But, if you lived here, you really didn't need all that much. It's been ideal.

Figure 53. Narrator John Baum: "I don't want to make a big deal out of this, but the Presbyterian Church was run by the salaried employees." (Photo: *Zinc Magazine;* Courtesy of Ogdensburg Historical Society)

John Naisby, another salaried employee of the Zinc Company, came to Franklin in 1937. Although he confirms that living in Franklin during the Depression had its advantages, he also verifies the low wage scale.

> When I moved up here in 1937, these people didn't even know there was a Depression on. They were still working five days a week and everything was going along just like usual. I mean, it was amazing. Anybody coming from where I came from, this was Utopia. My salary was ninety dollars a month so that was considered pretty good at that time.
>
> But it was hard to get any raises out of them. We generally didn't get any raise until the union arranged with the company to give the hourly employees more money. Then, maybe, the salaried fellows would get some more money. The way they got around paying you more money, give you a title. When I left, I was chief of construction and maintenance, I think it was.
>
> The Neighborhood House was the social center for

everyone. They always had activities there. We had bowling teams and we would bowl once a week. They had a library. And good housing, they were the "better homes." We called them the Bitter Homes. Your house cost you twenty-five dollars a month rent. Then in one section, they had the bungalows. In Ogdensburg, they had a whole bunch of bungalows. They were very cheap too.

The real estate end of it was very closely controlled. They were pretty strict. When McCann was in charge, he ruled the roost pretty well. I mean, he had his deputies in different places that could report all that was going on. They used to, years ago, have a fellow up in the tower there on the coarse crusher building that could look right down on the railroad to the main office. He was not very well liked by most of the people. He was kind of employed as more or less of a snooper.

Our house had a terrace in back. So I planted a flowerbed around the top of the terrace. And this guy who was in charge of the real estate, he came around one day. He says, "I don't think Mr. McCann would approve of your flower bed on the top of the terrace there." I said, "I don't know why not, but if he doesn't approve of it, let him tell me about it." But this is kind of stuff you had to be very careful about. They had these snoopers going around and then they'd report to Mr. McCann.

Eddie Mindlin lends a unique perspective to the Mining Oral History Project, as he is the only Jewish narrator. Here, he discusses the various recreational facilities, his involvement with the fire department, of which he is still a member, and his experience with discrimination.

The Neighborhood House or Community Club was the best thing about growing up in Franklin. [It was] sponsored by the Zinc Company. Do you remember the band concerts in the park? We used to always go to the band concerts. That was something. It used to be on Thursday nights. The band, the award-winning band of Franklin, was sponsored by the Zinc Company. They practiced at the Neighborhood House.

And swimming at the pond thanks to the Zinc Company. The Zinc Company supplied the lifeguards. All those lifeguards were Zinc Company. They did everything. The band. The pond. The safety. The hospital. They made a hospital second to none.

Figure 54. Franklin Firemen with their 1915 American La France Fire Truck, which was donated by the Zinc Company. (Photo: Courtesy of Franklin Historical Society)

The Zinc Company started the golf course. That was a unique, private club. The Wallkill Country Club. They gave this group of guys, mostly bosses at the Zinc Company, this property for the use of a golf course only and if it ever is anything else, it goes back to the Zinc Company. Sure, they wrote it off. Business is business.

We had a fire department. We had the first motorized pumper in the county [donated by the Zinc Company]. It was a 1915 American LaFrance. We still have it. Franklin, because of the Zinc Company, was the most equipped, modern, best in everything.

Sure. Sure. [There was discrimination], but very little. There was a nationality problem. If you were a Pollock, you couldn't join the fire department. Or a Russian. It wasn't only a Jew. The first one that got in that was not English or Irish or something like that, I was told was Kornheiser. He was a well-liked fellow. Julius Jacobson was number two. I became a fireman in 1956, I think. Thirty-one years.

You want to know about discrimination? When I graduated from high school in 1947 — after coming home from the navy — I went to the Zinc Company. I asked Truck Paddock for a job. I filled out the application and everything. Then he told me, "Well, all the summer jobs are filled and we're not hiring anymore." It was a week or within a week, Joe Beni got a job. His father, his brothers were working there. But I thought it [discrimination] was very little. You have to understand that the Zinc Company kept Franklin alive. Franklin, during the Depression, was one of the very, very few towns to have people working. The Zinc Company never laid off their men. They kept all their men. They worked them two days a week, three days a week. They had the smallest breadline. They had the smallest welfare group in any town around. That's how the Zinc Company took care of them.

Many Frankliners say, "Thank God for the Zinc Company." Others say, "Well, we could have had M&M [Mars Candy Company] here, if it wasn't for the Zinc Company." Do you remember in the forties when M&M wanted to come, but they couldn't get enough signatures for working people? The big fight against them was the Zinc Company. M&M was paying people the same price that they were paying people down in the hole, in the mine. They [the miners] weren't paid good. They had to work hard. Some of those miners were excellent workers. You couldn't beat them. The Zinc Company, as far as I'm concerned rating from one to ten, I'd rate them eight or nine. Sure, they restricted a lot of things, but they gave an awful lot to the community.

Like Eddie Mindlin, attorney Don Kovach feels the advantages of living in the company-owned community outweighed the disadvantages. Here he speaks of Police Chief Irons and the way the Zinc Company exerted control over its employees not only during working hours but also in their private lives.

There was one enforcer everybody was afraid of: Chief Irons. I was afraid of him as well. I had some run-ins with him. He was the enforcement for the Zinc Company. Things were a lot different. You didn't get to go out of the community very much. They had a judge. And they didn't have lawyers to speak of in those days. So, if you were brought on the carpet for something, it was handled locally. Locally meant that if there was something that was going to impact in any way on the Zinc Company, the result would be Zinc Company oriented.

I remember seeing some old correspondence, that was left behind in the Zinc Company main office, which as you know, I eventually bought and raised my family there. On the second and third floors was our home, and the ground floor became my law office. Anyway, I found some communications, old carbon copies, tissue type, which was from the nurse who the Zinc Company hired. The nurse was reporting on three or four employees, miners, who did not report to work. It was her conclusion that the reason that they didn't report [for work] was that they had gotten drunk, and that there was a drinking problem that the Zinc Company would have to focus on. Now, when you have only one job and there's no place to go and you can't speak English too good, if you get drunk on Sunday night and don't come to work on a Monday, and Monday afternoon the nurse comes visits you and before the day is out the personnel officer comes to visit you and says, "If you don't come to work Tuesday," or "If you do that again, you're fired," I think that kind of discipline probably worked pretty well. But, if you did your job and stayed healthy and didn't get hurt, most of the families that I knew had a good life. Families were a strong unit or aspect of the Franklin community. So I would say that the Zinc Company's paternalistic attitude was probably positive for most of those involved.

It was a very parochial, paternal type environment, but not one which was an unhappy experience. Most people were able to gain much more than they could have if they had not come from Europe. They had their own homes. They were able to buy them through wage deductions. We had most of the cultural things that larger communities have. We had recreational facilities. We had a good school system. All of which emanated from one capitalistic support and that was the New Jersey Zinc Company. But, they weren't benevolent. They were watching out for the Zinc Company.

Was the New Jersey Zinc Company a good employer as far as employees go? Well, I would have to qualify that to some extent depending upon where you were in the hierarchy. I think for the first generation immigrants, the work was very hard and the pay was not all that good. However, in comparison to what they would have been subjected to if they had not come here and gone to work for the Zinc Company, [it was good]. I'm not so sure [in Europe] that they would have been able to do, for example, what the next generation was able to do, which I believe I'm an example of. And this is to move on to higher and better things.

Although the United Mine Workers eventually was able to bring the union to the mines in Franklin and Ogdensburg, the union was organized through the Chemical Workers branch and, for the most part, was considered to be ineffective. Once Zinc Company management saw that a union no longer could be avoided, they courted the union representatives. Also, after Social Security came into existence in 1935, the Zinc Company changed their pension plan to reflect the benefits an employee would receive under the government plan, thus lowering the pension that would be received from the Zinc Company. According to former Sterling Hill employee, Wasco Hadowanetz, the union:

> . . . must have come in around the 1950s, but it wasn't very effective. They would go out on strike and get a ten-cent raise, say. At one point, they signed a contract that they would get a pension — the difference between a hundred dollars and Social Security. It was a very unfortunate thing because later on Social Security went up [but the Zinc Company pension did not go up]. They never raised that one hundred dollars. There were some that retired with a five-dollar pension from the mine. It was a very poor arrangement.
>
> The union people were catered to by the mine officials in those days and so the benefits weren't that great. They tried to keep on the right side [of the union representatives] but in doing so, those people that were in charge of the union weren't fighting very hard. They gave them easier jobs, those that were the union representatives. I heard that the union presidents and the representatives were getting soft jobs. And then, they would have a union hall up here in which they would make sure they had a couple kegs of beer. They got very little benefits. They'd go out on strike for a year and get a ten-cent raise. To me, there's something wrong there with the dangerous work that they were doing.

In Franklin, former Zinc Company employee Bill May, who worked in the company store from 1946 until it closed in 1951, also feels that the union did not accomplish much for the miners. As for the union calling a strike, Bill observes:

> I was salary. Everybody that worked in the store was salary. But, they tried that. Maybe that was after the war. I think that helped to split the men into factions, union, non-union. I don't know if that helped the company to close up sooner or not. I can remember these guys, and they looked like union men. Come in with a big black LaSalle, dressed up with the fedoras on and so on. All dressed up. These were the fellows that were trying to sway the men. They picked out different ones in the mines that they wanted to be union. Pro-union and get more money, "We'll get this," and "We'll get that." And they had the union bosses in both mines. And they would have a big beer party and try to get members to join. I think my father joined because it was a case of "You better join!" You know, you're in or you're out! Then, later on, he got out of it. He said he just couldn't see their way of thinking. It wasn't good. It was like the John L. Lewis type of thing, the labor bosses. I couldn't see it because of the type of people that were in the union. They looked like rabble-rousers or whatever. Even the ones in town. The fellows that were pushing the union — I can't say undesirables — but they were the ones they could get. They could really give them a big line and suck them into this way of doing things and so on. I couldn't see it.
>
> They went on strike one time. I forget how long it was. They ended up on the short end. The Zinc Company really gave it to them. They got ten cents an hour. The Zinc Company took away Saturday work. They got the shaft. You talk about ten cents an hour, eighty cents a day, five days a week, four dollars, right? They took away Saturday time and a half, which could have been, who knows, fifteen, twenty, thirty dollars. They gave them four dollars and took away twenty-five or something. I thought they figured, "You think you're going to give us the business, we're running the show here!" And that's how they retaliated against them. Hit you in the pocketbook, that's where it hurts. It didn't bother me because at that time I wasn't working for them, and if you were a salary man you couldn't be in the union anyway. You were a company man, that's what it was. That's why they gave you the position. All the bosses and everything were company men, salary men.

At one time, it was one of the richest zinc mines in the world, plus having all the other minerals in it. Out of every ton of ore, the zinc involved in it was like 40 percent. Out of a hundred tons, forty tons [were zinc], which was unreal. They had a mine in West Virginia, the yield was 3 percent and they were making money. So you can see what the Zinc Company made here in the early years of the mining operation. That must have been a multi-million dollar operation.

They owned everything. That was the case in a mining town. They had the homes. They had the water company. They provided the power from the power plant. It was all Zinc Company. Sure, they provided services for the employees, sure. There was pro and con. There was good points. They had all this stuff that was beneficial to the people in the town. Ogdensburg had the same thing in the Sterling Mine. They provided well.

We had the Neighborhood House downtown. That was run by the Zinc Company. They had their own staff in there. We had basketball at night at the high school. In the summertime, we had baseball that was run by the recreation department. The Franklin Pond where I spent my summers. I used to live down at the pond. They had skating in the wintertime. They had spotlights on the pond. You don't have that anymore. This day and age, things should get better instead of worse. And they do nothing down at the pond now. Nobody's picked up the slack where they left off. The town, that I can see, doesn't provide anything. Before it was available to everybody in town, and practically everybody in town was connected with the Zinc Company. Now, everything is gone.

Most of the narrators voice regrets that living conditions have deteriorated since the closing of the mine. Bob Svecz, who was born in Franklin in 1952, just two years before the Franklin Mine closed, eventually went to work as a miner at Sterling Hill in Ogdensburg. Bob continued to work at Sterling Hill until it closed in 1986. Although he never experienced the golden years of mining in Franklin, he is glad to have been a part of this unique mining history. Today, Bob works as a guide for the Sterling Hill

Mining Museum. He expresses pride in having been a part of Franklin, Sterling Hill, and the New Jersey Zinc Company.

> I've lived all of my life in Franklin. But, with the Zinc Company going, it seemed to take everything that the town had along with it, because I guess the company provided for everything. The Zinc Company owned the town. A lot of people have a lot of pros and cons for the Zinc Company. They did a lot of good and a lot of things they did just like any other corporation does that runs everything. They ran the town.
>
> I had heard how great things were. You asked me what it was like growing up — well, from hearing how things were before and when I'm growing up, it seemed like there was nothing compared to what they had before. Neighbors. Parents. Friends. Relatives. All told what it was like. You had the Neighborhood House. You had bowling alleys in the basement there. You had the Franklin Pond. We didn't even have a town park. Schuster Park had been closed. It was very important. Then when the mine closed, it went out like a . . . nobody cared. Yet it was a very important part of not only the history of Franklin and Ogdensburg, but for the entire state of New Jersey. Really. I did find it to be fascinating.

~

Figure 55. The Franklin Pond and the bathhouses built by the Zinc Company. (Photo: Carrie Papa)

Even though the Stonega Company, Calumet & Hecla, the Zinc Company, and other large paternalistic mining companies were good to their employees, the following incident related by Bob Shelton shows the pervasive authority involved in company ownership. By having company officials in key community positions, the Zinc Company exerted immense decision-making power. As director of the Sussex County Trust Company, and as an involved member of the community, Bob Shelton served on the school board along with Clarence Haight, mine superintendent of the New Jersey Zinc Company. In addition to being chairman of the school board, Haight also served as superintendent of the Presbyterian Church Sunday school for some twenty years. In his interview, Bob tells a little-known story about the Zinc Company houses that were given to the local school district in the late 1940s.

> I knew all of the [Zinc Company] officials. The mine captain and I served on the school board. Clarence Haight. He was a graduate of Columbia University. Smart man. The school board's a job. It's the toughest job in the world. It's a headache. The Zinc Company built eight houses [on School Plaza across from the school] to help get teachers because we couldn't get people up here. There was no place for teachers to live. The Zinc Company built them and gave them to the school district. That was part of their giving to the town. So the school district owned every one of those. And, of course, you found in them the [school] superintendent, the principal of the high school, the principal of the grammar school, and the school secretary.
>
> We had a meeting — and by now the Zinc Company is going, going and taking their taxes with them [Franklin will no longer have revenue from Zinc Company taxes]. So we had a principal leave and a new man came in and he said, yes, he would take the job. He's talking to the board. But they had to improve the house. Paint it. Improve the inside. I hit the table. BANG! "I move we sell the house!" Pandemonium! This is positively horrible. There is the high school superintendent sitting across from me. He lived in one of those houses. I couldn't get a second to that motion.
>
> I went home and I sat down and I figured out what those houses were costing us. The Zinc Company built those in

the old days because there was no place for teachers to live. But now we were only charging twenty-five dollars a month rent. Ideal location. They could walk to work. Now, we've got fifty teachers and we've got seven of them living in there. All for a long time. I made up this program of what it was costing us to keep those houses, and had the girl mimeograph it in the bank, and I sent one to everyone on the board. Haight was the board chairman. He was smart and he wanted to be right. He stopped me in the post office and he said, "Bob, I read those figures. This is terrible." I said, "Clarence, we're being taken and it's costing every tax payer in this town." He says, "I agree with you." I said to myself, "If he agrees with me, the rest of those board members are all Zinc Company men, everyone of them except Morris Herzenberg, they'll go along with it." So we sold the houses!

We sold those houses and Hollobaugh, who was the superintendent, never spoke to me again. Purdy, who was the high school super, was smart. He bought the superintendent's house. [laughter] He knew he was getting a steal. And Haight and I, we ended up very good friends.

The Zinc Company did a lot for the town. The better homes, they built all of those. They owned them and they sold them to their employees. Very reasonable prices.

We made loans to the miners. A lot of them bought their homes up there [in Siberia]. They were little homes. They're all improved now. Many of them very nice homes. But they were little four room shacks then. Had a privy out in the back. All that's gone. If they couldn't buy it, or didn't want to buy it, the rent was very cheap. They never charged a lot. They were not in the real estate business. They wanted to keep their operations going. And, in many ways, they were ahead of many of the companies.

The Zinc Company, by and large, was a pretty good outfit to deal with. They wanted to make a profit and they did. And anybody that bought their stock made money.

The bosses bought stock. English were buying stock. And you would be surprised how many of these miners who could not read or write bought stock. Our experience with these foreigners was very good. We loaned money to many of those people. They could hardly speak English, but they were hard working people and honest. The miners, we had very little trouble with those people.

Instead of feeling exploited or oppressed, most of the miners are grateful for the benefits that the company provided. This was true for other miners in other mining communities. While they may have had complaints about the companies they worked for, most miners and their families still felt proud of their communities and their way of life. Whatever their attitude about the New Jersey Zinc Company, most of those who grew up in Franklin or Ogdensburg would agree with narrator Genevieve Smith when she says:

> I think we've always known here in town that this was a unique place because of the mineral content in the mine. We've always been aware of that. Everybody just remembers the town with a great deal of warmth. I suppose [the sense of pride] was just the fact of growing up here, the town was part of me and I was part of it.

This sentiment applies to the residents of many towns in many mining areas, as well as New Jersey. Or, as Florence McCoy in *Appalachian Coal Mining Memories* sums up life in the coal fields of Virginia's New River Valley:

> Everybody lived just about the same. And we were happy.[7]

---

1. US President's Commission on Coal, *The American Coal Miner*, 33.

2. *A Medical Survey of the Bituminous Coal Industry*.

3. Garner, ed. *The Company Town: Architecture and Society in the Early Industrial Age*, 187.

4. Thurner, *Calumet Copper and People*, 8, 36.

5. Raymond L. McCann joined the New Jersey Zinc Company in 1917. He held a degree in mining engineering from Lehigh University. In 1939, McCann was named general manager of the Franklin and Ogdensburg mines. He was elected president of the company in 1951 and chairman of the board in 1966.

6. The New Jersey Zinc Company Pension Department was established Jan.1, 1911.

7. LaLone, ed. *Appalachian Coal Mining Memories*, 68.

Brick Row, Siberia, Little Mexico, and Hunk Town

America as the melting pot of the world was truly exemplified Saturday when The New Jersey Zinc Company held its annual picnic in Franklin's attractive Shuster Park. There the native born Americans, the English, Scotch, Irish, Italians, Russians, Poles, Hungarians, Czecho-Slovakians, Germans, Austrians, Spaniards, Chileans, and Mexicans congregated to the number of several thousand.

Zinc Magazine, October 1924

Among the many goods, benefits, and services the New Jersey Zinc Company provided for its employees were the houses it built and maintained. Company housing in the various neighborhoods of Franklin and Ogdensburg ranged from small bungalows for the miners to the truly elegant home on Main Street provided for the mine superintendent. The neighborhood where one lived was determined by ethnic background and the position held within the company — an ethnic and economic distinction that was not unique to the mining industry. It was tacitly understood by mine owners and miners alike that immigrants preferred to be with others of their own nationality and, as a result, immigrants were settled in segregated neighborhoods. The majority of miners, especially immigrants, seemed to accept the placement in ethnic neighborhoods with very little resentment. In fact, some appear to have taken life in those communities with a sense of gratitude rather than hostility. The ethnic neighborhoods in Franklin were known as "Brick Row," inhabited primarily by the English; "Siberia," where various Eastern Europeans lived; and "Little Mexico," a district for married laborers from Mexico and South America. In Ogdensburg, there was only one ethnic section of town and that was known as "Hunk Town." As was typical of most mining communities, the Zinc Company also maintained boarding houses for single professionals and immigrant laborers.

Former geologist John Baum explains the structure and cost of the homes supplied by the Zinc Company for its employees.

> The Zinc Company ran a boarding house for the bachelors. This was next to the main office on Main Street in Franklin, the old Catlin house. I boarded there for a year. It was run by a lady and she had a small staff to look out for the place. We were sent off with our lunch buckets after breakfast every morning and they provided dinners at night. The cost of the room, single room, was ten dollars a month. If you shared a room with someone, it was five dollars. Room and board, I believe, were thirty dollars. That may not sound like much money, but it happened to be a large proportion of what we made. My pay at that time was five dollars a day. From that pay, which was something like fifty-five dollars, Social Security was deducted and our health benefits, things of that sort. After a year's apprenticeship with the Zinc Company as a bachelor, I married and the Zinc Company supplied me with a home to rent. Ultimately, I bought the house for $3,350. This was a house I lived in for twenty years. What they call the better homes. Up in Rutherford Avenue and Green Street, Master Street, that whole area. The company had fifty homes there they built for salaried men, primarily, shift bosses and the like.
>
> Of course this town was fully stratified. I won't say we had a class system, but as I told you, the home that I lived in, fifty of them were built for salary people. Other homes were built in quantity for hourly workers in an area called Siberia, up on the hill to the east of Route 23. A section called Mexico was just off Sterling Street, west of Main Street. That was built for hourly workers.
>
> Then there were better homes throughout the town, such as Fowler Street. Large, magnificent homes in those days. Some of them two family homes. They were built for salaried

Figure 56. Narrator John Baum said that the homes on Catlin Circle in back of the main office of the Zinc Company were built for super department heads. (Photo: *Zinc Magazine;* Courtesy of Ogdensburg Historical Society)

people. Department heads. Of course there was a home built for the superintendent, on top of the hill here. Now, these are stratified. No two ways about it.

We had the Jews and the Catholics, all lumped together with the foreigners. Among the young bachelor engineers, it was said the only eligible girls were teachers or nurses. It didn't always work out that way, naturally. [Prejudice and discrimination] were not as strong as I give the impression of because McCann, the superintendent here, knew everybody by name. He was friendly with everyone.

Robert Metsger, who was Sterling Hill's superintendent when it closed in 1986, also speaks of housing in Franklin and confirms the existence of class distinctions that John calls stratified and Robert terms a caste system.

The reason I lived in Sparta is because when I first came in 1949 and was looking for a place to live, there weren't any houses available. The Zinc Company had houses, but they were all filled at the time. Hunting around, I found it was a

Figure 57. Close up of the "Better Homes" on Green Street and Rutherford Avenue. (Photo: Carrie Papa)

lot cheaper to buy an old summerhouse that had been winterized down at Lake Mohawk than it was to rent an apartment up in Beaver Lake. But there was a real caste system in Franklin that I never liked. There was the management. The geologists were invited to parties given by those people. Then there were the engineers, the surveyors. They were professional people, but they were not part of that group. I don't know what it was. Then there were others. I guess the miners were another group.

But I was never really a part of it since I didn't live in Franklin. [The miners] were good people. I mean, the Europeans, the middle Europeans maybe, were not able to make much money, but their children were all brought up well, mostly. In my work, I would talk with them and I saw them all the time. Knew them and they knew me. But I was not their boss or anything like that.

Alan Tillison, who grew up in Hamburg and did not start working for the Zinc Company until after he was married, says:

We rented when we were first married. The house on High Street was a company house. They didn't overcharge you.

Figure 58. Zinc Company cottages built for the miners in "Siberia" as they appear today. (Photo: Carrie Papa)

> Whenever you rented, anyplace, they did the maintenance. At the time, real estate was real cheap. They started selling the Zinc Company houses, like those little bungalows and the better homes down there. They started selling those in later years. You could buy one of those little bungalows for about twelve, fifteen hundred dollars. But you didn't have twelve or fifteen hundred dollars! [laughter] Plus, the better homes, I think they sold for around thirty-five hundred or thirty-seven hundred. But that was a lot of money at that time, when you were working for sixty-three cents an hour.

Steve Revay, whose father worked in the mine and who himself started working for the Zinc Company in 1922, remembers the company house he lived in as a child.

> Portables. They were hooked on an engine. There was a lot of them portables around for awhile. When the Zinc Company started, they were the first buildings they brought in for people to live in. We lived in a portable bungalow. It was tough living. All outside toilets. No running water. All that stuff. One night, a big, heavy storm come up and blew

the roof right off of the damn building. We had to go to the neighbors and stay overnight.

In the beginning, we didn't have no electricity. Water you had to go out to get. You'd go get your water. Greenspot. That was on Main Street. Between the Neighborhood House and the house across the street. We used to get the water there, but when they sunk the shaft and started the mine, that disappeared.

That's one good thing the Zinc Company did. After they built all these homes, they sold them to the employees. Siberia. There's about two or three hundred there. The bungalows were good at the time. They were better than the portables we lived in. We moved out of the portables into these bungalows. They sold them for nine, ten hundred dollars. You'd pay ten percent down and then they'd take it out of your pay.

And they built the better homes. These had bathtubs in them. They had good sinks, good toilets, good kitchen cabinets, and everything like that. We called them the better homes. The Hunkies, they got nothing, you know. When they built all those bungalows up here, they didn't put no bathrooms in them. All they put in was the toilets. Rusty iron toilets and iron sinks. Our home was cold. Oh, it was cold. You never got warmed.

We've been in this house sixty-one years. You know what this house was? A Mexican restaurant. They brought a lot of Mexicans here to work in the mine and they didn't have no place to eat, so some guy opened a restaurant here, and they used to eat here. It was a close knit community. You had good times.

In the following, Franklin miner Nick Zipco notes how the different ethnic backgrounds of employees determined their positions in the mine.

From England, they're the ones were the bosses. But who made the mine? The Russians and the Pollacks and the Hungarians. They're the ones made the mine. They're the miners. They were the real miners. The bosses were English. I only seen one boss really work. I heard he was a boss, but he was working like a miner, but he wasn't no miner. He couldn't do it. He was slow. Slow! But these Hungarians and Russians, they produced. We had the Slovaks who produced. I went in a place in the pillar when it was dinnertime. This is dinnertime, you're supposed to be at lunch. Here's this guy pulling these handles, pulling the ore. I couldn't see him. Had to get right on top of him to see him and he was eating his lunch. He wanted to make a bonus. Work all the time. Certain ones did that. No overtime, but it was all set what you had to do. Seven and a half tons a day, that's what you had to produce before you got anything. After that, bonus. We had the Russians and the Mexicans here a lot. Siberia was where the Russians lived, up there in company houses and the Mexicans were in the Mexico area on this side. Sterling Street. We got along pretty good. We couldn't complain. [Discrimination] — there wasn't none. Everybody got along good down in the mine. There wasn't no enemies like.

While former Franklin miner Alan Tillison agreed with Nick Zipco that most of the miners came from Eastern Europe and most of the bosses from England, he at first did not believe that this was discrimination. Only after some reflection did Alan admit that bias or prejudice could account for the fact that the majority of the bosses were English.

There weren't too many "native" Americans working there. There were mostly Russian, Slavic, Polish, Hungarian and English. Most of the English were all born in England, and they'd been miners in England. They came here and worked in the mine. They had the supervisory jobs. I don't think it was discrimination. They had the expertise. Later on, it could be a little bit of discrimination too, because you had your higher echelon that were all the same type of people. I think of all the shift foremen — shift bosses you call them — in the mine, there was only one that wasn't English.

Tom Sliker worked in the Ogdensburg mine from 1941 until the mining operations were shut down in 1986. Tom, whose background is German, remembers those of different nationalities who worked in the mine and lived in the various ethnic neighborhoods.

They had Siberia. They had the Russian section. They had a Mexican section. And the Slovaks had their own little section up in Siberia. But most of the people I worked with were Englishmen. The people came from Cornwall. My wife came from Cornwall. That's where my father-in-law came from. Came here and started mining here. He came by himself. His family stayed in Cornwall until he got established here and then he brought them out. Others came here and mined also. He brought Joe Weeks who became his son-in-law after he brought him out here. Cousin Jacks. You don't hear that term anymore. But, I got along with the Hungarians and the English. We had a lot of Hungarian bosses in the mine. And we had Mexican men working with us. All nice people. I never had anything against any of the nationalities.

Bill May, who worked in the Zinc Company store, describes housing in Franklin, and discusses his friends and the various ethnic neighborhoods where they lived.

I was born and raised in Franklin, down on Church Street by the Catholic Church. My grandfather, Bill Jones, the Zinc Company brought him here from Michigan to sink the Palmer Shaft, which was the beginning of the Zinc Company. My father worked for the Zinc Company for thirty-five years. We were middle-class. They used to call up by the golf course Nob Hill. Nob or Snob, both, I guess they referred to it. Like the four hundred were years ago. The elite or whatever. You had the Pelletts, who was the chief engineer. Then you had his brother who was the town dentist. You had the mayor, Mayor Jenkins at the time. Don McKechnie, the plant superintendent. They all lived up there.

Then you had a place over here called Siberia. Siberia, why? Because you had a lot of foreigners living in there. You had a place called Mexico over on the other side of town. Things like that, you know. There was discrimination.

In the mine, practically all the foremen were Englishmen because of their knowledge or whatever. That probably was discrimination too. Hell, they said at one time here, years ago in town, the English and the Irish ran the place. If you weren't English or Irish, you were out. I got in the fire department when I was twenty-one. At that time, there was nobody in the fire department other than English and Irish.

Figure 59. Another section of town was known as Brick Row or Michigan Row because of the number of people from Michigan who lived there. (Photo: Carrie Papa)

Nobody else. I went in with the first Jewish guy in the fire department. I think that was 1951. That was more or less the breaking up of the clique. Yeah. Then you started getting in the Polish guys and maybe Hungarian or Italian or whatever. But that was it. That was strange. I never realized that at the time, but I look back at it now. Julius Jacobson. We went in the fire department together. Up until that point, there wasn't any Jewish in it. Oh, sure, there was discrimination.

I went to school with Polish, Carl Zucknovich; Hungarian, Freddie Kotan; Italian, Natalie Lorenzo. Her father had the garage in town for years. Ann Trofimuk, Russian. Jean and Jane Houyoux, twins, were French. It was like a cross section of Europe in this mining town, here.

In fact, they used to bring, in cattle cars up from Mexico, Mexican labor years ago. They had their own development in Ogdensburg and they had it in Franklin too. They had big boarding houses. There were a couple. There was a girl I went to school with, her father was a Peruvian. Peruvian was tin mining. I don't know how they got word of it, but some Peruvians came up here too. People stayed on here. I went to school with a boy named Garcia. He was a Mexican. Maybe some of them made money and went back because if you took money back there, they would have been big shots. Compared to around here. But there was discrimination too. They had their own place in town, which wasn't the greatest. The housing was just housing. Whereas, you had in Franklin here, up around Green Street, all the better homes. The two story homes were all the supervisors and the bosses.

I can remember, years ago, my father had chickens for awhile. Everybody had that. We were going to sell the eggs. Couldn't give them away. Of course, the people that lived up by the golf course didn't have to worry about raising their own chickens or having a garden. If they had anything, it might have been flower beds or something. I used to go up there. A friend of mine, his name was Hilton Brown. The name Hilton sound familiar? His uncle was Conrad Hilton, the multi-millionaire hotel baron. Sure. His mother was Conrad Hilton's sister. The name Brown was from the father, Connie Brown his name was. He was the electrical engineer for the Zinc Company. Jackson Pellett, I was very friendly with their boy. I used to stay with him in a big home up there on the hill. Anything you touched in that house was money, even an ashtray. He was an inventor and he was the chief engineer for the Zinc Company. Jackson Pellett, we used to call him Stonewall Jackson.

In fact, his wife was Lieutenant Rowan's daughter. Lieutenant Rowan. I can remember reading in school a story called *Message to Garcia*, in the Spanish American War [Lieutenant Rowan was the hero of the story]. He was Bob Pellet's grandfather. His mother's father, Lieutenant Rowan, who also became an inventor. I went to his place in New York when I was a kid. He had the penthouse on top of the Pennsylvania Hotel. This place was filled with all of his inventions and stuff, little working models. His name was Rowan. Yup. Bob was my friend. We went in with Mrs. Pellett. She went in to buy art supplies. She was an artist. We had lunch in this fancy restaurant. There must have been twelve pieces of silverware and Bob says, "Just watch my mother." This fork and that fork and do this and do that. Oh, Jeez. But very nice people. [We remained friends.] Bob is in Maine now. He married one of these French girls I told you about, Jean Houyoux.

I also went to school with Donnie McKechnie. His father was the plant superintendent. I used to play with him in baseball, football, basketball. We were on the team at school at the time. The fellows I mentioned, they were on the team too. Carl Zucknovich. In fact, he was a big star in sports. Freddie Kotar, the Hungarian boy, he was on there. Steve Vargo and Ray Heredia, who was, I think, possibly Mexican.

It was good. They [the Zinc Company] provided well. That was the case in a mining town. They owned everything. They had the homes. They had the water company. They provided the power from the power plant. It was all Zinc Company. It was to provide services. They had all this stuff that was beneficial to the people in the town. Ogdensburg, the same thing in the Sterling Mine. It was good. I liked it.

Ann Trofimuk, the daughter of a Russian miner, experienced the same type of ethnic, economic, and cultural mixing of school friends. She reports on several of her friendships.

I was friendly with Mary Lou Jenkins. She was one of my best friends. Miner's daughter and the mayor's daughter. A lovely lasting friendship. We haven't seen each other in years, but we go way back. We were a duo. We sang in school together. I mean, it was just great.

Another one of my best friends, also in high school — because Mary Lou was taken out of school and sent to a prep school in Virginia — was Ruth Garris. Her father was a bigwig in the Zinc Company in Ogdensburg. In school, we just were best friends. [My cousin] Pete's good friend was McCann's son. Mr. McCann used to take Pete and his son — also Earl Smith's son Jack — to New York to ball games. Earl Smith was in charge of maintenance, the carpentry, and the painting, this kind of thing.

Going back to generations before, they do say there was discrimination. They might have said "Hunkies" or whatever. But I don't remember that. I do remember Mickey McSavage, he's deceased now, but I remember his telling me that years ago when he was working in the mine, there was something. He was not given a particular job because he was of Russian background. That it went to somebody who was what they would call a "Cousin Jack," the English people.

As far as my own experience goes, we had food. We had shelter. I guess my feelings were that the Zinc Company was there and it took care of things. I mean, it was the paternal type of thing. We felt it was a marvelous time to grow up in

Figure 60. Narrator Ann Trofimuk [left] and friends represent four nationalities — Russian, Czechoslovakian, Polish, and Irish — all living and playing together in the same neighborhood. (Photo: From the author's collection)

> Franklin. There were so many special things. Maybe we forget the rough edges, but it was just a safe time and a happy time.

Even though Wasco Hadowanetz lived in Hunk Town as a Zinc Company employee, he did not feel a sense of discrimination. He points out:

> There was an ethnic division in town. Down on Avenue "A" and Avenue "B," that was referred to as Hunk Town, because the Eastern European miners mostly were living there. There were Slovaks, Ukrainians, as I am, Hungarians. Actually, they had Mexicans recruited from Mexico and Texas, and they lived down in that area too.
>
> I grew up in a company house. During those years, for the eleven dollars we were paying for rent, they would fix your roof or paint the house periodically. Of course, we had outhouses, and they cleaned those out. They would bring slabs of wood for you to burn in wood stoves. I bought the house after my father died. [Living in Hunk Town] didn't

seem to be a derisive thing to me. We knew that many of us were poor down in that area, but it didn't seem that there were any that were wealthy. In other words, there was a division, but I don't recall that there was any animosity per se. I remember some of the fellows from Sterling Hill would be coming down the hill and we would have snowball fights. But we all played marbles in the yard and I don't remember any kind of derision.

There were company houses on Avenue "A," Avenue "B," Plant Street, and Bridge Street. Bridge Street was known as the "Dump" because that's where they dumped the tailings from the ore, and then they built houses on top of them. The supervisors usually lived up on Sterling Hill. That's up behind the mine.

Chilean immigrant John Pavia, who started working as a miner at Sterling Hill in 1927, eventually bought a Zinc Company house on Bridge Street. When he first lived in the house, John reports:

> [The rent was] fifteen bucks. Fifteen dollars. I bought this one from the company. I think it was forty-five dollars a month. Quite a few years, they took out of my pay. I think I bought it in the fifties. I can't remember.
>
> My wife, she never liked down here [on Bridge Street, also called The Dump]. She was from up on Sterling Hill, you know. Yup. She never liked it down here. Up there was mostly all foremen. You know. Mackey, he was a big wheel up there. He used to holler at my wife because she used to rollerskate by his house. She used to live right across from him. And when she would go to school, she used to rollerskate. Mackey. He was Mackey the Terrible. [laughter] He was a big boss in the mine. There were a lot of bosses in the mine [living up there]. Up on Sterling Hill, the houses the company built, they was the better homes, they called them. For the engineers and the foremen.
>
> For the working people, the company built all the houses. This one. Next door. That one. The next one down. I think there was eight on this street. This one here was built in 1927. Then two houses over there are maybe a hundred years old. Then on the other side too, all the other houses; they're all the company. But my wife, she never liked it down here.

Don Kovach first became aware of ethnic and economic differences when he entered the Franklin school system. He says:

> There was a distinct caste system that was maybe necessary. I've never harbored any adversity or blame, but there definitely was a difference. I learned that as I went through the school system, and after I also had a stint in the mine. My family would tell me that the immigrants from Eastern Europe were imported, if you will, for purposes of doing the hard labor. They were hearty men and were able to go into the mines and do what was necessary to satisfy the requirements of hard rock mining of zinc.
>
> The bosses, until later, were all English. Many from England, who had been coal miners in Cornwall or the Cornish area. The reason for that was from a practical standpoint. They could read and write [English]. They could keep the necessary records, communicate with the officers and directors, were simply able to do the administrative end of what was necessary. Whereas, the immigrants were relegated to lesser jobs. But that was okay because they had an opportunity that they never would have had in Europe, with most of them coming from a peasant background.
>
> As a matter of fact, I went back to Hungary one time when my dad was still alive. It was in 1976, and, even at that point in time, it was easy to see that they left because there's poverty there. It was obvious that fifty, sixty, seventy years before that, it must have been terrible. It was still the Austria-Hungarian Empire and they were actually serfs.
>
> But anyway, there was a discrimination that ran throughout the community. Yes, there was, in my opinion, a definite discrimination type thing, but not anywhere near what other minorities have gone through. And the fact that we became educated and contributed and went on to achieve dissipated that relatively early on, I would say, but it did exist.

As a Jew, Eddie Mindlin experienced bigotry on a personal level, yet he still retained an appreciation for his hometown of Franklin, and its benefactor, the New Jersey Zinc Company. Eddie acknowledges the discrimination:

> ... but I thought it was very little. You know, this Jew bastard and stuff like that, that goes on all over. That was

always around. No matter where I've gone, I've always said that I was Jewish first because that's what I was and that's what I am. And, if you're going to be saying remarks about the Jews, it lets you know that I'm hearing it. So you watch yourself. You don't have to talk that way. There's a lot of people that don't even mean it. It's just a figure of speech when they say, "Oh, that guy, he's a cheap Jew." It's just a figure of speech.

About this prejudice thing, in 1936, Dr. Vermes came to Franklin. Left Europe before the War, came to Franklin. He heard about a Hungarian town, with a lot of Hungarians so this is where he came. He opened up his first place in a building that was owned by my uncle. He moved in and opened up his first practice. Isadore Toppal advised him, he says, "If you want to make out in this town, which you will because you're Hungarian and they need a doctor, don't claim you're Jewish. Don't say anything." So he didn't say anything and he joined the Catholic Church because he should be affiliated with a church. He was excellent with a violin. His wife was an accomplished pianist and they put on some duets in the church and all of the different organizations. They were excellent musicians.

He was Jewish — Vermes — you know. He died and he's buried in Israel. There's a Jewish saying, "Cut me up in little pieces, but throw me among my own." So no matter how far he got away from Judaism — he was a pillar in the Catholic Church, Hungarian-Catholic Church — but he went back to his own when he died.

His wife Alice never said she wasn't Jewish after he died and she was alone. They always donated to the synagogue. To get established, that's what a lot of Jewish had to do, deny their faith. Bill Goldstein changed his name to Bill Gray.

You knew these things were there. You expected them and you either worked around it or you let it go. My wife asked me one time, she said, "I don't understand," she said, "I never knew a Jew that acts like you." Well, I'm raised among Gentiles, I didn't know how Jewish people live.

Figure 61. Narrator Eddie Mindlin remembers Dr. Vermes, who belonged to this church. (Photo: *Zinc Magazine;* Courtesy of Ogdensburg Historical Society)

Whatever experience of discrimination people living in Franklin might have had, the majority seems to feel as Eddie does when he says:

> I've always liked Franklin. Franklin was my favorite home. That's one of the reasons why I never left. My friends were all here. They never treated me any different. The little discrimination that I got from this town wouldn't chase me away.

Can't Make Ends Meet

The stock market crash of 1929 started a downward spiral in prices, production, employment, and foreign trade. Collapse in commodity prices reduced buying power everywhere and increased unemployment in all industrialized countries. Before the end of 1929 the entire economy began to snowball downhill. Every kind of business suffered and had to discharge employees; they, unable to find other jobs, defaulted installment payments and exhausted their savings to live. America had neither Social Security nor unemployment insurance.

The Oxford History of the American People

After the stock market crash of 1929, Americans reeled from the economic crisis that affected their daily lives. As banks and businesses failed, factories and mines closed, farm prices fell to unheard-of lows, and millions of people lost their jobs, life-altering changes were forced upon suffering citizens. Three of these changes that were most often noted by the narrators in the Mining Oral History Project were the shifting roles of women, the dramatic cuts in salary and hours that mining company employees endured, and the increasing use of credit at local stores and the company store. Of course, as is evident in *Appalachian Coal Mining Memories* and *Hard Times*, mining families across the nation, not just in Franklin and Ogdensburg, experienced these same sociological traumas that resulted from the country's economic crisis.

In the midst of the Great Depression and in other times of economic hardship, women worked along with their husbands to maintain the family. They were responsible for such money saving efforts as gardening and canning food, raising chickens, selling eggs, picking berries, and making clothes. Women in poverty-stricken areas often set the children to work gathering nuts and berries, picking up coal and wood for heating, or selling and delivering milk or newspapers. Money earned in this way became part of the family income. Several narrators in the Mining Oral History Project remember their mothers' roles in the household during the Great Depression.

Another life-changing result of the economic hardship was that many mining companies, including the New Jersey Zinc Company, cut back on wages and hours to keep their men working. In oral histories, government surveys, and other historical writings, men report that they worked in the mines only one or two days a week. In *Coal Towns,* Crandall Shifflett summarizes the wage cuts of several Appalachian mining companies. At Wheelwright, Kentucky, the Inland Steel Company cut wages from over six dollars a day to two dollars and eighty cents during the Depression. In Michigan, Calumet & Hecla laid off all single men without dependents, and instituted a rotating system of work for those miners who remained employed. The night shift was abolished and wages were reduced by 20 percent. Many mining companies also made efforts to contribute to the welfare of employees and their families. More than once it was recorded that company detectives looked the other way when women and children gathered coal that had fallen from the company trains.[1]

The use of credit at stores, known as being "on the book," became a common occurrence during the Great Depression and one to which many narrators referred. Although this seemed to be the most convenient practice for shoppers who did not carry cash, exploitation was inherent in the system. An example of this, as many historians point out, is that some mining companies began to pay the employees' salaries in script, company-printed money redeemable only in the company store. The Zinc Company store did accept credit, but the company neither issued script nor forced employees to shop in its store. None of the narrators spoke of coercion or cheating by the Zinc Company.

Many of the narrators in the Mining Oral History Project speak of the life changes they experienced in the era of the Great Depression but, at the same time, express appreciation for the small pleasures they were able to retain during those difficult times.

Here Laura Falcone remembers growing up in a period of economic distress.

My parents came to Franklin during the Depression, in the thirties, and it definitely was because Franklin was a place to earn some money. My father was very accepting. He was willing to do his day's work without complaining and never managing to make ends meet. That was one of the big terms I remember in my family, "Can't make ends meet." With four children during the Depression, I can understand that now as an adult.

I remember Daddy's hat — there was a special name for it. It had this carbon light on the front of it and it smelled funny. Very sharp, smoke smell. He would bring it home and take it back with him. It wasn't left at the mine as far as I know. I remember boots. He had to have heavy boots. It would be a hot day out and he'd say, "I have to take my boots. They're working in a wet area." And it would be down in these mines and it always frightened me. But Daddy never, to my knowledge, complained of any problems. But Daddy was sick for a long time, and I think this contributed to their being in debt all of the time. There might have been hospital insurance, but this would not have covered everything. Daddy being sick is probably what put my parents in such a bad state in the end. I don't know if the company had a health plan that would pay people if they were not working. I don't think it was typical in those days.

It was a company town, but I don't think I knew this when we were little and living there. But, when Tennessee Ernie Ford sang or recorded that song *Sixteen Tons* about the coal miners and they owed their soul to the company store, you then remembered that the company store would give credit. And, of course, we always had a bill at the company store. Well, they'd get it paid off. Then they'd owe again because the store was a general store that sold food, clothing, everything. There were other small department stores in town that we would buy from too. Some were a little less expensive than the company store, or if we owed too much at the company store, we would then have to go to another store to buy things until we got paid, or whatever the problem was. My sister Evie was a baby, I believe, when we moved to Franklin. We never lived in a house that had hot water running. We were always in like what was called in the city, tenements, cold water flats. That's what we had. That's what most people had.

I remember when I was fourteen, which had to be 1944, I took over the family finances. I had all of the bills in a little pot in the china closet. I would try to have them in order and pay the proper one at the proper time. There was no discretionary money left over. Mom was happy to have me take over. It was a horrible job because there was never enough to go around. But, regardless, Mom insisted that my kid sister, Evie, have a little fun in her life, so she went to the canteen, which was in the Neighborhood House during the war. It cost fifty cents to go to the canteen, and you could buy two pounds of hamburger for fifty cents, which would make a magnificent Sunday meal! I always fought with my mother about Evie's fifty cents. Not a strong dispute. I mean, Evie went to the canteen. Meat was a treat. I mean, hamburger, not even meat. A roast, I don't ever remember eating a roast at home.

One thing Mom made a lot, and this would be our entire meal, was a big thick pancake made in a frying pan, probably fried in lard. It was about an inch thick with an absolutely delicious crisp, crunchy outside. Mom would slice it in triangles like a pie, and she would sprinkle sugar on it. We all loved it. We also had a lot of fried potatoes, again probably fried in lard. Other than these two fried items, we had fairly healthy food. Rice and grain, which are the best things you can eat today. Red meat, my goodness, today it's bad news.

We had a lot of vegetables. Everybody had a garden. Everybody. During the war they called them Victory Gardens. If you didn't have property on which to grow a garden, the town provided land for you. That was over on Buckwheat Road. Empty land that was owned by the town, I suppose, and anyone that wanted space could have a garden there. I remember our neighbors, who lived in an upstairs apartment without a yard, going over to Buckwheat Road to their garden. But the house we were in had space in the back. Mom and Dad worked in the garden. All week we'd pick things. And Mom did a lot of canning. And making jelly. We'd go berry picking and Mom would make jelly. She

Figure 62. Narrator Laura Falcone: "I was down at the Franklin Pond watching the kids skate." (Photo: From the author's collection)

worked very hard. I suppose we were very poor, but we never realized it at the time.

In fact, I remember the first time I found out that not everyone was poor, or lived like us. Every Christmas, Mom and Dad would send to Sears and Roebuck for a Christmas order, some new clothes, a few toys maybe, and a five-pound box of chocolates. That five-pound box of candy probably was the most thrilling thing about Christmas, all of that candy! How we loved those chocolates. Well, with four little girls, I don't imagine that box of candy lasted very long. At any rate, I remember, it was always gone before we went back to school after the Christmas vacation. Well, one day in February, I was down at the Franklin Pond watching the kids skate. We didn't have ice skates at that time. And, for whatever reason, my friend Marguerite was there and invited me to go home with her for awhile. She lived up on what I later learned was called Snob Hill, the better part of town where the Zinc Company officials lived and so on. At that time I didn't know there were different sections of town, or that there was any distinction at all between one kid and another.

I remember the house was big, but that didn't mean anything to me then. While we were at her house, Marguerite offered me a piece of candy from this big box of chocolates, most likely a five-pound box. I remember I was surprised that they still had candy a month after Christmas. But then, when she opened the box, I was amazed that they were still on the top layer! That anyone could still have chocolates this long after Christmas was astounding, and so much of it that they didn't even get to the second layer yet! This just amazed me. Why, they must be able to have candy whenever they wanted it! Not just at Christmas.

I never knew until that moment that everybody didn't live the way we did. Once I realized that Marguerite could eat chocolate at any time, I knew there was a difference between us, between being rich and poor. Until then, I don't think I had any inkling at all that we were poor. Nearly everyone's father was connected with the mine in some way, and I would guess we just never thought about who was a laborer and who was a professional. Certainly not in the younger years. Later on, in high school, we must have been aware of social and economic differences, but I don't think it bothered us.

Mom made a lot of our clothes. Hand-me-downs. That's all we ever lived in. Four girls, it was very convenient to keep passing things down. But one year I had my own new dress. I remember my mother sitting up all one night making an Easter dress, a yellow dimity Easter dress for me. The dimity was a beautiful pale yellow with tiny bright yellow flowers and a stripe woven into the material. I thought it was the most beautiful thing in the world. I can recall walking down that big hill from High Street to the Presbyterian Church for the Easter Sunrise Service. That is such a clear memory and such a happy one.

My father died in 1951, but probably the year before he died, there was a big strike at the mine. There was no such thing as insurance, which is what probably put my parents in such a bad state at the end. But I'll never forget the day the strike was over, or the company had decided to give in to whatever was required. One of the things that they settled was that the miners would all get a twenty-five hundred dollar life insurance policy. Daddy was sick in bed in the house on Nestor Street. The man came to the door and said he had to sign to be joining this program. And Mom said, "He's not well. He's bad." And the man said, "It doesn't matter, he just could put an X on it." Well, it seems to me, within months Daddy was dead. And Mom, in spite of losing her husband, was able to pay off everything that they owed. If it had not been for that man coming and Daddy signing the paper, she would never have been able to pay her bills.

It sounds like a sad, hard life for Daddy, but I think he was a happy man. A delightful man. He took enjoyment wherever he could find it, and saw the humor in life. When I was fourteen, I became interested in going to the Presbyterian Church. It might have been one of my girl friends in school that encouraged me. I think it was. At any rate, I was to be baptized and my father dressed up, and my mother too, and they came to the church to see it. And that was the very day that the minister announced his retirement, or he was moving on to somewhere else, or something. Whatever, he was leaving. And Daddy always joked, "I went to church for the first time and drove the minister away." [laughter] Happy. Find fun. That was Daddy. My sister Nora had the same talent. Enjoy every aspect of life. Always find fun. And I think my kid sister Evie has the same talent.

Evelyn Sabo, the kid sister "Evie" of whom Laura spoke, recalls additional childhood memories and confirms that her father did have a great capacity for enjoying life, even if life was unendingly harsh.

I think it was a very difficult life. I know it was hard work. He had to go down the mine shaft everyday. He would wear this miner's cap that had the lantern in front. He would tell us the stories of mining ore down there. But he had a lot of friends that were miners also, and he would talk about the good times. A lot of them were immigrants, which were of Hungarian descent, Polish descent, and he would say a few words now and then in these different languages because he was very quick in picking up their languages and conversing with them.

I think he enjoyed the fellows he worked with. I can't say that he enjoyed the work. It was very hard work. He would always come home tired and dirty. I don't think he was a very strong man to do this hard physical labor. But I never heard him complain about it. As far as health problems go, he ended up with cancer and he died at a very early age, forty-eight. [While he was sick] I don't think there was any compensation, or if he got compensation, it couldn't have been very much because they were always struggling. I remember when we moved from the High Street house, we moved into a house on Main Street. But we only lived there

Figure 63. Narrator Evelyn Sabo. "They were always struggling. . . . but . . . I didn't realize we were counted as poor. Very poor." (Photo: Courtesy of Evelyn Sabo)

for a year because they couldn't afford the coal that it took to heat the house. We had to move into a lower rent house that was much smaller.

I never ate in a restaurant. [We had] potato soup. Things that were low cost. I didn't know steak in those days or a lot of the better cuts of meat like lamb chops or anything like that. If I had any luxuries like a stick of gum, I was only allowed a half a stick because it had to be shared. If I had a penny candy, I would buy root beer drops only because they would last so long. I also remember that I almost never got an ice cream cone because it cost five cents.

One day, Laura's friend Caroline, she and I, for some reason, were going downtown. She was going to treat me to an ice cream cone. I was so thrilled I was going to get a whole nickel ice cream cone to myself. I probably was maybe eight or nine years old at this time. So we went into Herzenberg's drug store where they had a soda fountain. She bought me the ice cream cone and we were walking back and I'm licking it and the ice cream fell off the cone. I was devastated! I was crying because I lost my nickel ice cream. Anyway, she didn't have money to buy me another and I was just out of luck. But I was crushed by it.

I also remember one time when Mom was trying to make some money. She would make doughnuts to sell and she'd make fudge to sell. At this time, I was probably about five years old and we were allowed to eat some of the doughnut holes, but we could never eat a doughnut. Also, I would go with my sisters because they were older. We went up by the school where they had the homes up there, for the teachers that worked at the school, to sell the fudge and the doughnuts up there. The one door my sister knocked at, it was one of the school teachers and so she sold her some of the fudge and I guess the school teacher felt so sorry for me — I probably looked so sad and so small — she gave me a penny. I couldn't imagine how anyone could be so generous and give me that penny. She said, "You can buy anything you want with it," and I was just thrilled to get that penny.

I remember, to entertain myself I would draw paper dolls and cut them out and create little dresses and outfits for them with tabs on them so I could put them on the dolls. I spent a lot of time playing with paper dolls. I also remember egg crates because all the eggs used to come in a little egg crate thing and I used to pretend that was all post office boxes and write notes to all pretend people and receive notes from pretend people. I remember taking clothes pins and putting them together [two pins together] like an "X" and putting them all around the floor creating my beautiful home with so many rooms, and I'd wander through all these rooms.

Every once in a while we'd get a box from welfare. At that time, I didn't realize it was welfare. I remember there was one doll in there, kind of raggedy now I recall, but at that time just to have a doll all my own . . . . I remember calling her Marie. I loved the name Marie. Because of the Marie doll, when I had my first daughter, I called her middle name Marie because of that doll that I loved so much.

We had a very meager existence really, but at that time as a child, I didn't realize we were counted as poor. Very poor.

Actually, few of the miner's children were cognizant of their poverty during the Depression. Since nearly everyone around them was in the same situation, most children managed to enjoy their days without being aware of the austerity of their lives. Former Sterling Hill employee Wasco Hadowanetz, whose father worked in the mine during the Depression, describes several childhood jobs he held and wonders how his parents were able to cope as well as they did.

In the thirties — I was born in 1930 — and by that time we had seven in the family. Then we had ten in the family. I really don't know how my parents managed. In the early days during the Depression, my father was making ten, fifteen, twenty cents an hour. The Zinc Company, to their credit, didn't lay off anybody during the Depression. They cut down to two days a week or so, but they never laid anybody off.

When I was growing up one of the chores we had was to go up in the woods — we had a little wagon that my father made — we had to chop down trees to have wood for the winter. Then, later on, we had coal. We'd have to sift the coal out after it was burned to get the "clinkers" [unburned coal] out and reuse that coal. My father, he was pretty strict. We had all of our chores to do and they had to be done. There was no sassing back, or "I'm not going to do it." There couldn't be a lot of disobedience. Everybody had to know his job.

When you look at our family, only two of us went through high school. What happened was, they usually quit after eighth grade and either went to work or had to work at home. After grammar school, we just thought we were going to quit and go to work. I remember there was a Mr. Zabrisky who had a farm in Ogdensburg, and my brothers used to pick tomatoes there during the summer.

As a youngster, I would go down to Lake Mohawk and caddy at the golf course. Lake Mohawk was an exclusive area. They had the gatehouse. In fact, in those days I don't think they allowed Jews or Italians to build there. It was really exclusive. These people were from New York and New Jersey suburbs and they had summer homes there. Played golf, had boats on the lake. The golf course is just off Lake Mohawk. Many young boys went there and caddied. It was one of the few places where you could make any money.

One day, I was walking along the street on my way to the golf course and a milkman stopped me. He said, "Would you like a job delivering milk?" So for the next five years or so, I peddled milk. I helped him load the truck. In those days it was not homogenized milk. We had grade "A," grade "B," and Guernsey milk with more cream on the top. Then we would sell eggs and I'd have to candle the eggs to see which ones were good and bad. We sold butter. Every house was of a different design with different levels. Some of them, you had to walk up and put milk on their porch. Others, you put it down by the bottom steps. All these things you had to remember. Then he had me doing these other odd jobs; I had to sink a well in his yard in Sparta. It was four foot in diameter. He would lift the tiles down and then I'd dig around underneath until they would drop down, until water level at about twenty-five feet was reached. Imagine being inside a four-foot hole, standing in cold water. But that's how it was. We all worked.

Like Wasco, Franklin resident Steve Revay thinks back to the Depression and the need to gather fuel.

During the Depression, the Zinc Company run three days a week. Of course, every place else was shut down completely almost. But, if you worked for the Zinc Company and you worked three days a week, the money that you got for them three days wasn't enough to live on.

I used to have to go up and dig coal up there on Beaver Lake Mountain. They had a train house up there where the engines used to load up with coal and there would be a lot of soft coal spilled all over the place, you know. We used to go up there and pick it up and bring it home to keep warm in the winter. When I was a kid, our home was cold. Damn, it was cold.

Now, if you worked for the Zinc Company for three days, hell, you couldn't afford to buy coal. You couldn't afford to buy food almost. It wasn't enough, but I worked the three days a week.

Eighty-five-year-old Nick Zipco remembers the difficulties of his younger years. Nick came to Franklin with his family when he was just twelve years old. By the time he was fifteen, his mother and father were separated, and as Nick says:

We had a really hard time in life. I tried to find a job in the worst way. When I was fifteen years old, I started in the silk mill weaving. I had to learn first. I worked nine years weaving before I went into the mine. I worked in the silk mill and I took care of the family. Seven children. Four girls and three boys. My mother was a good mother until we lost everything in Pennsylvania. [After the separation] I was the oldest [child] at home at the time. They wanted to split our family.

This social worker came six o'clock in the morning in my house and wanted to take all my sisters away from us. [They were young] and she wanted to take me too. My father didn't want the kids to go away from our home and I didn't want them to go away from our home. I didn't care what. She came many times to the house, but this time she come six o'clock in the morning and I was furious.

They took one of my sisters when I wasn't home. I come home and I'm looking for my sister, "Where is she?" "Oh, they took her away." They took her to a home where a lady wanted to keep her, a private home. And the next day she was back. You know that kid cried all night, and come home the next day. I didn't want to see my sisters split up and not know where they are. My youngest brother, he had a hard life. He was a little kid when he was milking cows.

We lived in the Lacey Houses. There was seven houses in one place. We didn't have running water. We had to get it from the roof. Rain water. I worked on this one [my present house] all my life. My wife was worried where I was because I was here lots of times at twelve o'clock at night. After I worked so many years, I got two weeks vacation and I stayed home and worked on the house. For two years, we didn't have no heat in this house. We had peach baskets in here [as furniture] first two years. She had a little baby. Every day, she'd go uptown. No refrigerator. Every day uptown and spend one dollar every day.

In the following, Ann Trofimuk, who worked at the company store during her high school years, remembers the changes in both shopping and banking that resulted from the economic hardship in Franklin.

I have a dim recollection, when we bought stuff from Toppal [local merchant] way back, and from Koller [local merchant] they had like a brown covered book and anything you bought, anything, they would write out the date and what you bought and how much it cost. Then, at the end, maybe it was a week, you went and you paid. I remember I would have to check the addition and everything before the bill was paid. I would say most people bought on the book.

I remember going to the bank, where somewhere along the line, they must have set it up for me to go cash my father's check. You can't do it in today's world, right? But, I remember, going to the bank with my father. Up that High Street hill and down and sitting on those marvelous marble benches [in the bank] with my legs dangling, so I must have been pretty young. Pop always went to Mr. Shelton. However they set this up that I was going to cash Pop's check for him, I would go to Mr. Shelton. I was little so he obviously had seen me in line. I always remember him sort of peering over his counter — my father's name was John — and saying, "Well, John, how are things down on the three hundred level?" And I thought, "Oh, my goodness." But, here again, how safe it was. I don't remember, but that money must have been put in an envelope or something and my trudging home with it.

I have a dim recollection of Mr. Mindlin, of him standing on the corner there where the lights are. Whether that was to catch people and get paid, or maybe somebody was delinquent. But, to my knowledge, my parents always paid on time. There wasn't any of that business of their being behind and having to be reminded. I imagine that, as in all of life, that people had problems where they couldn't pay on time, and maybe there was this business where they [the merchants] felt it was wise to catch the miners when they got paid.

Bob Shelton, former director of the Sussex County Trust Company, which was the only bank in town at that time, explains how it was possible for Ann to cash her father's checks. Additionally, Bob clarifies "book accounts" and talks about merchants cashing checks for the miners on pay day.

The miners could come up here and cash their checks. They didn't even have to sign their name. They had bearer checks. So if somebody took that check and brought it up, they could cash it. But we knew those men. If somebody came in [that shouldn't have a check] we'd say, "What the hell are you doing with that check?" And the merchants would come in and get two thousand dollars or three thousand dollars on a one-day note. They'd sign a note. In order to cash the checks. The stores. The miners would go in the stores and pay the bill they owed. Because many of them carried book accounts as they called them. Many of them couldn't read or write. The merchants carried accounts. Most of the merchants were fair with them. Then, the next day, they'd bring the checks in and we would give them a credit to their accounts. Pay off the note. So it worked out fine. We never really had any trouble with that system. And none of the merchants that we could tell of were dishonest. There were none where we had to say, "We won't lend you the money."

The Zinc Company worked through the Depression. They were good customers of ours. They wanted a bank here for their employees. Many of them had stock in our bank. They didn't control the bank by stock holding. But they had the biggest account. They had several accounts. And they were very helpful to the people here. The majority of people that worked as miners had accounts in the bank.

Eddie Mindlin's father, who owned a store, cashed checks in both Franklin and Ogdensburg by going to the mine gate on payday to see the miners as they left work. On his father's banking practices, Eddie reports:

> To go back to Zinc Company days, my father had to go on the credit business too because they [the company store] gave credit. Ogdensburg and Franklin was like a brother and sister town. My father had customers in the Franklin Mine and he had many, many customers in the Ogdensburg mine. Franklin was paid every other Friday. Ogdensburg was paid every other Thursday. So Pop used to go to the Franklin Bank, see Mr. Munson and shake hands. "I'd like to borrow five hundred dollars." Then, go cash checks [for the miners as they left work].
>
> They [the miners] used to pay my father two dollars a payday. Three dollars a payday. Whatever he could get, but the only way to get that money before they spent it, was to cash their checks. He used to stand outside the time house in front of the Zinc Company store. My father used to beat them [the other merchants]. Others did it too. Whoever ran credit. Isadore Toppal had his area, the Sterling Street and upper Main Street area.
>
> Then, after he cashed the checks, he used to pay back his loan and, with the extra money, [collected from the miners] he would deposit it. That's the way it was. The stores wanted to do it because they could get their payments, [before the money was spent on something else].

Former Zinc Company geologist and present curator of the Franklin Mineral Museum, John Baum, discloses additional information about the credit system used by the company store and the other merchants.

> The whole town was built by the company. But there were merchants in the town who were not company-dominated, you might say. In the beginning, they charged exorbitant prices to the point where the company built a general store in order to make things available to the miners at fair rates. That's the opposite of what you hear about the company store and mining towns. This company store, well, I won't say it was charity, but it was of great benefit to the miners. They could run up credit there. They could run over their credit there. Of course, the company store was going to get this money out of their salaries one way or another. But they had prices that were cheaper than the rest. They kept the merchant's prices down, and did a very nice job doing it. So we did our shopping in town, or in my family, much of it was done by mail order to Sears Roebuck.
>
> [The miners] shopped in the company store and with the other merchants as well, who would extend them credit. It was interesting, on paydays every two weeks, to see the merchants lined up, outside the time office where the miners came out of the plant. They had large sums of money ready to cash their checks, to be sure of getting their money. Catch the miners as they came out with their paychecks in their hands. It was also interesting to see a number of wives standing there to catch the employees coming out with a paycheck because the wives knew the first thing those men are going to do is head for the bar and spend some of their money. So the wives grabbed it. [laughter]
>
> They [the Zinc Company] did not pay terribly well, but all during the Depression people had jobs from the mine when they did not elsewhere in the county. Over in Newton, they used to say if somebody pulled a twenty dollar bill out of his pocket, they knew he was from Franklin. They also said over in Newton, that if they saw anybody walking in the street over there, they knew they were from Franklin because Franklin didn't have sidewalks. [laughter]

Tom Sliker, who was hired as a mucker at Sterling Hill but ended his working career as the mine superintendent, started working at an early age. Unlike many of the other narrators,

Figure 64. Main Street in Franklin was still unpaved and had no side-walks during the Depression years. (Photo: From the author's collection)

however, Tom was quite aware of his early poverty. The hard-ships during Tom's formative childhood years influenced his entire life. Tom reveals:

> When I was born, I was born a real poor person. And I mean poor, extra poor. And I was poor until I started two jobs. I started farming when I was about nine years old. Where I lived, there was two Englishmen that had a farm. The old man died and the lady asked me if I would come there mornings and nights after school and help with the farm. So I stayed there until I was nineteen. Then I got married and moved to Franklin.
>
> Five years later, I got a job in the New Jersey Zinc Company in Ogdensburg. There was an opening there. 1941. April the first, I started working in the New Jersey Zinc Company. I started as a mucker. At the same time I'm working in the company here, I had a painting project job for myself. I was contracting jobs out on a second job to make enough money to live on. I was a painter and a paperhanger. I had my wife and two children. I painted for thirty years in a second job. When I was working for the company, some weeks I'd work the afternoon shift. Then I'd work daytime painting. If I worked the day shift, I'd come home and go to work at three-thirty painting until nine at night. Saturdays and Sundays, I always had a job. I always wanted something for my kids that I never had.
>
> So I always thought for them. I never went to basketball games. I never went bowling. I never went to the bar room.

Anywhere's I went, I went with my wife and my children. And I saved my money as I went along, all the years I worked in the mine. My first day of work, I started at fifty-six cents an hour. Then I went to fifty-eight cents an hour to seventy, eighty cents an hour.

Then, when I became salary, I made a little over two thousand a month. Twenty-three hundred a month. That was in 1966. I was salary. When I quit at the age of seventy, I had run the mine for twenty years. It was a good life. The New Jersey Zinc Company, they were good. All way round, they were good.

During the years of the Depression, the Zinc Company continued to mine zinc even though it could not sell it at the time. One of the narrators reports that a huge pile of zinc ore was kept in the Zinc Company yard: "It was during the Depression. Pile of Ore. They just put it in there until the Depression was over and then they sold it. They cleaned it out after. But they had a big pile right in the yard there."

Although Robert Metsger, geologist and last superintendent at Sterling Hill, did not join the Zinc Company until 1949, long after the Depression, he understands that the New Jersey Zinc Company had practiced a sound long-term economic policy. Bob explains:

> They were conservative in saving up when times were good, saving the money for when times were bad. The Zinc Company used to stash its profits away for a rainy day when they could do a lot of exploration. One of the things about mining is that there are times when it is a bust. For a mining company to have a lot of money in the treasury is a good thing because when times are good, they can make a lot of money. When times are bad and there's no call for their ore, and they can't make much money, and things in general are cheaper because the bust times are bad, the company can then use the money in its treasury to explore and develop new properties. That secures its future.
>
> I understand, and of course this is hearsay, but I understand that during the Depression, for example, there wasn't anyone laid off at Franklin. The men were cut down to maybe a couple of days, but people always had a job.

That's another thing — something you don't find in companies today — security in your job. I may not have made as much money while working for the Zinc Company, but I had a job continuously for forty years.

The feelings of the majority of people who worked for the New Jersey Zinc Company, especially those who worked during the Depression, are reflected in the words of Clarence Case. Both Clarence and his father worked in the Sterling Hill Mine for a number of years. Proud of being a miner, Clarence proclaims:

> If I had it to do over again, I would have done it again. Oh, certainly! Absolutely! When I was growing up most everybody in Ogdensburg worked for the Zinc Company. The New Jersey Zinc Company — they can say what they want about the company — I know that my dad worked during the Depression days. He worked three days a week where other companies were laying off. That says a lot. It was a great company to work for.

---

1. Shifflett, *Coal Towns*, 108.

Health Care and the Phyllis Club

The Franklin Hospital, the first in Sussex County, was erected in 1908 by the New Jersey Zinc Company to furnish prompt aid to the company's employees and in addition, to supply hospital service to residents of Sussex County.

Franklin Borough Golden Jubilee

At the turn of the century, health care in mining towns, as well as in the rest of rural America, was limited. The isolation of many mining communities made satisfactory medical care difficult to obtain. However, while inadequate, the majority of mining companies did make some attempt to furnish at least minimum medical care. Most mining companies kept a doctor on the payroll, but all too often he served several mines, traveled long distances, and did not have the training or facilities with which to provide specialized care. Some mining operations, however, provided only a nurse or a dispensary, and difficult or serious cases requiring more advanced care were transported, sometimes great distances, to the nearest hospital.

In contrast to minimum care, some of the big corporations in the mining industry such as Stonega Coke and Coal Company in Virginia, Colorado Fuel & Iron Company, or Calumet & Hecla Company in Michigan provided first-rate medical programs. As early as 1868, Calumet & Hecla had a local hospital to serve its miners, who paid fifty cents a month if single or a dollar a month if married to help cover medical costs. The Cornish system of paying for medical and hospital care, a method initiated by immigrants in the 1840s, was used by most mining

203

companies. In this system both the miner and the company contributed to medical care. By the early 1900s, the Cornish health care plan was widely used throughout the United States.

At the time the Franklin Hospital was built in 1908, the New Jersey Zinc Company was one of the few mining companies to provide such a facility. The Franklin Hospital was directed by a surgeon-in-chief, who, along with the entire hospital staff, was hired by the Zinc Company. The first surgeon-in-chief was Dr. Frederick P. Wilbur, a graduate of Johns Hopkins University. Prior to the opening of the Franklin Hospital, all persons needing hospital treatment had been sent to Paterson, New Jersey, or Middletown, New York, and the long railroad journeys to these towns were sometimes detrimental to the patient's health.

Patients at the Franklin Hospital were charged seven dollars a week for ward beds and fifteen dollars a week for private rooms. This charge covered room, board, medicine, surgical dressings, and the general attendance of a nurse. Reflecting the ethics and morality of the period and the paternalism of the Zinc Company, the Franklin Hospital had strict rules for both patients and visitors. For example, patients were "forbidden to use profane or indecent language; to express immoral sentiments; to play at any game for money; to smoke tobacco in the house, or to procure for themselves or others any intoxicating liquors." Visitors were allowed only between the hours of two and three in the afternoon and "must not converse with any patient who is not relative or friend."

Although none of the narrators could recall being treated by Dr. Wilbur, they did remember many other chief surgeons. John Baum, who came to Franklin in 1939, knew Dr. James H. Spencer, who was surgeon-in-chief from 1934 to 1942, and his successor, Dr. William B. Boyd, Jr., who served from 1942 to 1953.

John recounts that his children were born in the hospital, a practice relatively unknown in rural American until after World War II. In 1958, a new wing was added on to the Franklin Hospital to provide a maternity unit. The new facility contained a labor room, four beds, eight bassinets, toilet facilities, and a formula

Figure 65. The Franklin Hospital was built by the Zinc Company. (Photo: *Zinc Magazine;* Courtesy of Ogdensburg Historical Society)

room. John discusses the cost for hospital-assisted childbirth and his memories of the hospital and staff.

> After I had been married a year and some, the first child was born and he was born at the hospital. The doctor's fee for the whole thing was fifty dollars. And I believe it was something like that for my second child. I had two children. Both boys, both born at the hospital and in each case delivered by the company doctor. Dr. Spencer was the company doctor when my first child was born. Dr. Boyd for the second.
>
> The Zinc Company built the hospital. If someone was injured, the doctor came. If he was injured in the mine, well, the doctor came down into the mine. The doctor was hired by the Zinc Company. These were good doctors. Surgeons. They took care of me. I got injured on the job. Not very seriously, but that was taken care of. The hospital service was open with very reasonable fees to all the employees. The Zinc Company, of course, took care of its own. If the miner got hurt, the company put him in the hospital and took care of him. Of course if a miner got sick or flu or something like that, that was on him. He just lost time from work. There was

unemployment insurance through the state. But if anybody got sick in those days, they got sick and paid the doctor themselves. This was just natural throughout the nation as far as I know. Dr. Spencer was a graduate of the University of Pennsylvania Medical School. Dr. Boyd was at Columbia. The hospital only had one chief doctor at a time, but other doctors in the area had the use of the hospital facilities.

Julia Novak, wife of narrator Steve Novak, was a registered nurse who worked in the hospital with Dr. Boyd. She recalls that he was a wonderful doctor, really "an all round doctor and one of the best surgeons." Steve agrees with Julia's evaluation of Dr. Boyd and supports her assessment with two incidents that display Dr. Boyd's character and reputation.

> Gerald Sharpe, he had some problem, I forget what it was, but he went to the city to see a doctor. The doctor said, "Where you from?" Gerald told him. "Don't you have a Dr. Boyd down there?" He said, "Yes." "Well, hell, you don't need me. You've got the best doctor in the country right there in Franklin." After I came back from the service, Dr. Boyd took me for a ride, and he wanted to know about my experiences. He wanted to go in the service himself and be a doctor because he knew he would get all kinds of cases. He was rejected because he had a lung problem, so they wouldn't take him. But he wanted so much to go into service because he knew he would get a terrific practice there.
>
> If somebody couldn't afford much, he would take care of them. One time, there was a drowning. He dove — it was cold weather — he dove right in the water to get the kid out. I saw him. Down at the Junction. Down at the Wallkill.

Genevieve Smith, who was employed by the New Jersey Zinc Company as a nurse at the Franklin Hospital for eighteen years, also remembers Dr. Boyd. Genevieve's father, grandfather, and grandmother were all employees of the New Jersey Zinc Company and all had a connection with health care in Franklin. Genevieve completed her nurses' training at Philadelphia General Hospital and had extensive military experience, serving as a nurse in North Africa, Sicily, and France. Here she

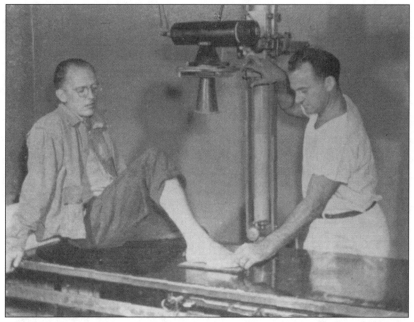

Figure 66. An employee of the Zinc Company has his foot x-rayed at the Franklin Hospital. (Photo: *Zinc Magazine;* Courtesy of Ogdensburg Historical Society)

discusses the Franklin Hospital staff and facilities, the health care provided by the Zinc Company, and the epidemics that affected the town.

> [The hospital] was excellent. I really can't praise the hospital enough. Now, the Franklin Hospital was not a large hospital. It was approximately forty beds. But it was excellent. Excellent! When I first went there, in 1946, the doctor in charge was Dr. William Boyd and then it was Dr. John Schmidt. The doctors were doctors that came out of New York, Columbia Presbyterian, and Cornell Medical Center. Dr. Schmidt had had extensive service experience, too, in the Pacific. They were excellent, excellent physicians.
>
> The [nursing] staff mostly were girls that had come from Franklin, Ogdensburg, Sussex, Sparta perhaps, from the area. They had gone away to school and come back here and worked in the hospital. It was a small hospital, but it had a wonderful staff and people were taken care of.
>
> The New Jersey Zinc Company had a nurse on staff at the mine so that when their men were out ill, she visited them.

Found out what had to be done and took care of that, saw that they got to the right place. Children and families, if they were families of company employees — and almost everybody in town was employed by the New Jersey Zinc Company — they were well taken care of. It was a paternalistic kind of thing. They were very good to their men. I really feel that they had very caring people running the company.

I know that there were companies with company towns like the mining towns in Appalachia, where the mine [owners] were not as careful of their men. This was a safe mine. They employed a whole safety division. Every accident was investigated. Steps were taken to prevent more accidents of that kind. They had a dispensary. The lab technician and the x-ray technicians that were in the hospital would take turns and, with the doctors, they would go to the dispensary from three to four every day. Anybody who had a boo-boo of any kind came into the dispensary. They did physicals on their men yearly. It was a well taken care of town.

My father drove the first ambulance in Sussex County. It was a horse and wagon and he used to take the men from the mine that needed hospitalization before we built our hospital, he had to take them down over Mase Mountain to Dover and

Figure 67. Narrator Genevieve Smith's father drove the first ambulance in Sussex County. (Photo: *Zinc Magazine;* Courtesy of Ogdensburg Historical Society)

Figure 68. Genevieve Smith's father also had been hired by the New Jersey Zinc Company as an apprentice harness maker for the mules that towed the mine cars back and forth between the mine tunnels and the surface. (Photo: Courtesy of the Bureau of Mines Collection; National Mine Health & Safety Academy)

get them on the train to get them into Paterson. My father worked on top. My dad didn't like the mine. My father was hired by the New Jersey Zinc Company as an apprentice harness maker for the mules that were down in the mine carrying the mine cars back and forth. My dad worked for them for fifty years, from 1907 to 1958.

At first, he worked in the barns that were down by the pond. And then, as soon as they got cars, he just took a car and learned how to drive it, how to take it apart and that sort of thing, and he was a chauffeur the rest of the time. The main office was in New York so he'd go down there.

We had a town public health nurse too and if you had chickenpox, or measles, or what have you, she came in to see you. Mrs. Glynn came and then on the outside of the house was a big yellow sign that says QUARANTINE.

I must have been about nine or ten when they had that
isolation hospital up in Siberia here. We had scarlet fever
here. All the kids were getting scarlet fever. And the company
took one of their houses that they had up in the Siberia section
up here and they turned it into an isolation hospital. They
took all of the kids that would have been quarantined with
the scarlet fever and put them up there. Now, my grandmother
and grandfather went there. Grandpa took care of the
maintenance and my grandma cooked for them while they
were up there. They both worked at the hospital too. This
was around 1930, 1932, I would say.

We always had epidemics of measles, mumps,
chickenpox. That one scarlet fever is the only one that I
remember. The typhoid epidemic, I think that came before I
was born. And there was a flu epidemic and they had used
the Neighborhood House at that time. That was the flu
epidemic during World War I. Phyllis Treloar, she was a
nurse, she was tied up with that flu epidemic. Her sister was
a nurse too, the school nurse. At any rate, Phyllis died during
the flu epidemic.

Alvah Davis, who started working for the Zinc Company
in 1922, remembers the flu epidemic, the use of the Neighborhood House as an auxiliary hospital, and the young nurse, Phyllis
Treloar. Miss Treloar was a student nurse who gave her life serving the community during the influenza epidemic of 1918. When
a women's club was organized in Franklin in 1920, they chose
the name of *The Phyllis Club* for their organization so as to honor
this young woman who had given so unselfishly of her time,
service, and life. Members of the club were committed to doing
charity work and to promoting good fellowship among the
women in the Franklin area. Alvah also reports on the health
plan that was offered by the Zinc Company at that time.

It was quite an epidemic. The Neighborhood House had
a lot of people into it. Sick. They took them out of the homes,
I guess, and put them in the different places. There was a
girl, a nurse by the name of Treloar, Phyllis Treloar — Did
you ever hear of The Phyllis Club? That originated at that
time. She was a nurse and nursed a lot of people. She died
nursing people and they named that club after her. It was

sort of a woman's club. My wife belonged to it. They used to meet once a month and they had a scholarship program.

You didn't have the health benefits that you have today. If I remember right, you paid a dollar a month. Then, if you got sick, you got twelve dollars a week. I don't remember how many weeks you would get that, but that's the only health plan that I remember. Of course, they built the hospital. My one daughter was born there. The others were born at home. But you didn't get many breaks with that [the hospital]. You paid your bill there. The doctor was twenty-five dollars. But I don't remember about the hospital.

Steve Revay, who was born in 1907, not only remembers the influenza epidemic of 1918 but also a typhoid epidemic in 1922 that had a profound effect on his life. Steve recollects:

I remember the war. They had a flu epidemic too at that time. That was quite a session too. They had that Neighborhood House full of beds. They used that as a hospital because the other one up there didn't have enough room. You know that hospital was built because there was a lot of accidents in the mine and they didn't have no place for them to go. Used to have to take them to Paterson. Down to St. Joseph. So, to save money, the Zinc Company built the hospital and they hired a doctor. Then, if somebody got hurt in the mine, they stayed here and went in to the Franklin Hospital. They didn't have to take them to Paterson.

Now, this is another story. After I graduated from grammar school, 1922, I went to Newton High School. See, Franklin had a vocational school. They didn't have a high school. If you wanted to go to high school, you had to go to Newton. So we used to go over to Newton on the train everyday. Well, while I was going to Newton every day, my mother got sick and my brother got sick. At that time, they didn't know what the hell was wrong with them. Well, I'm jumping ahead of my story here a little bit. First, let me explain. At that time the Franklin water pressure was very low because Franklin didn't have any water. When they sunk the mine, all the springs in town, that water all disappeared because of the mine. They didn't have no water.

In the meantime, the Zinc Company was trying to put water in the homes, here and there. There was a great big tank in the Franklin yard, the Zinc Company yard. The Zinc

Company built it for their own use if they had a fire in the plant anywhere. They had their tank and water. Well, you know what they did? They hooked that tank to the drinking water in Franklin and everybody says that tank was full of typhoid germs. So these typhoid germs got in the drinking water and it was a mess. In my family, there were three people that had it, my mother, my brother, and myself. So I was going to Newton and I got sick and they took us to the hospital. The typhoid fever, a lot of people died. I was in a room with five people and four of them died.

The hospital was owned by the Zinc Company. They took care of everybody that had typhoid fever. They caused it, but they never compensated anybody for anything. They took care of us, but they didn't compensate. Like me, what the hell, I didn't have no dealings with the Zinc Company at all. I got typhoid fever. I lost a year of school. I never went back because of that.

Although Eddie Mindlin was only three years old at the time of the scarlet fever epidemic, he was told that a nurse at the Neighborhood House actually gave him the name of Eddie during that epidemic. Eddie relates the story of his name and continues with details of how his brother started an ambulance service.

When they had that scarlet fever epidemic, I was three years old. I'm named after my father's father, Itchok, Issac. But my nickname was Kotchka. They called me Kotchka because I used to be short and fat and I wobbled like a duck and Kotchka is a duck. It could be Jewish. It could be Polish, Russian. So, [when they recorded his name for treatment at the Neighborhood House, during the epidemic], one of the nurses says, "That's not a name for a beautiful boy like that." She says, "I like the name Eddie." And my mother says, "All right, we'll call him Eddie." And that's how I got the name Eddie.

Let me tell you about the Lions Club because I think it's important because it was my brother David's idea. He wanted to join Kiwanis and start a free ambulance service. There was no [free] ambulance. By the way, you talk about ambulance and prejudice and stuff like that, as a matter of fact, it goes back to when my brother was blackballed from the firehouse. I was at that meeting. And the guy said that he only stood up and said it only as a joke, a gesture, but my brother failed to approve. That was Jack Ramsey, [who owned the funeral parlor]. Jack Ramsey was kind of perturbed at my brother. He had a right to be. Ramsey had got an ambulance and he charged for ambulance service. He was the only ambulance in Franklin.

That was after the war, the forties. My brother had an idea [to have this free ambulance service]. This was the original start of the Wallkill First Aid Squad. He wanted to join Kiwanis [to start this service]. There were Jews in Kiwanis, but they were limited to something like two of the same type businesses. My brother represented Sam Mindlin's Furniture. There was Barney's [furniture store]. The other was the upholstery store on Main Street. They wouldn't let him join Kiwanis because there was two furniture-affiliated businessmen there already. They could have bent the law if they wanted. I think they wanted to keep him out, [because they didn't want to get involved with the ambulance service].

Shortly afterwards, he was in Auggie Lorenzo's garage. Auggie Lorenzo was a friend. They were talking. This guy came in to buy a car. He was a big wheel in the Lions Club, and he asked, "Is there any Lions Club around here?" "No." He [Auggie] says, "There's Kiwanis, and there's the Elks in Sparta. There's American Legion and stuff, but there's no Lions." "Well, why don't you start a Lions Club?"

So they started it, my brother David, and Auggie Lorenzo, and this guy got charter members. Mike Stefkowich was a charter member. Bill Sabo was a charter member. I think we started the Lions Club in 1952. So my brother said his idea was to make a First Aid Squad. There was a guy, Phil Barish, had a Chrysler agency in Ogdensburg. Barish was a member of our Lions Club. Barish put up a Plymouth car to raise money. We were going to raffle the car off and with the money put the down payment on a ambulance. They started the First Aid Ambulance Squad. It was my brother's idea, and because of that — it was free ambulance and that put Ramsey out of business. He never liked my brother because of that. That's how come Ramsey got a little mad at my brother. He had it all [funeral parlor and ambulance service] and he gave up the ambulance [because of the free ambulance service of the First Aid Squad].

Laura Falcone remembers how frightening it was to see the ambulance in use. Even as late as the 1950s, doctors did not send patients to the hospital unless they were gravely ill. Laura recalls quarantine signs, some old-time remedies, and how the schools were beginning to teach health classes.

I remember seeing the ambulance arrive next door to take Mrs. Wilson to the hospital. It was so scary because nobody went to the hospital unless they were near death. We played with the kids and we didn't know what to say to them. We were sure Mrs. Wilson was dying, but actually she came home about a week later.

Usually, the doctor came to the house. But, there again, you didn't have the doctor come unless you were very sick. There was a nurse who came around. If you missed school, the truant officer was there right away, and if you were sick, then the nurse came. If you had a contagious disease — mumps, whooping cough, chicken pox, diphtheria, measles, scarlet fever, or anything like that — you were quarantined. The nurse put a sign on your door saying mumps, measles, or whatever, and you could not go out of the house until the sign was removed. I'm sure we had our share of childhood diseases and quarantines.

With four little kids, I would think if one got sick, the rest would catch it too. With colds, Mom would make us an onion poultice, which was cooked onions in a little cloth sack that we wore around our neck. I can't imagine what that was to accomplish. Vicks was rubbed on us too, and I also remember mustard plasters for something. While on the subject of health, I should say that I was the only one of four children to be born in a hospital. The rest were all born at home. Maybe there were complications with my birth, as it certainly was more common for children to be born in the home rather than in the hospital.

I remember health and hygiene were taught in the school. Tuberculosis was a great fear and there was a lot of school training about that. Especially about washing your hands and not spitting in the street. One time, they showed us a poster with twelve photos on it and we had to guess who had TB. The point was that you couldn't tell from looking at a person that they were sick. If you did have TB, you were sent away to a sanitarium. It was very scary. There was a family that had two little girls who had to be sent away. I was so scared of the disease I would hold my breath when I walked past their house so I wouldn't get any germs.

Poor Daddy was in and out of the hospital so much, and he still remained cheerful. I don't know who paid the hospital and doctor bills. Maybe they did have some kind of hospitalization because I do recall Daddy joking. He had to go back in the hospital for another operation and he said, "Well, I'm getting my money's worth out of the insurance company." Of course he was on a ward, not in a private room. But that was probably better because he was such a friendly person. As soon as he could move around from the operation, he got to know everyone. He would do little things for the ones that couldn't get out of bed, and keep people smiling.

Imagine though, we had a hospital in our tiny town. The town was two or three thousand, maybe. Yet, we had a fully equipped hospital. Of the age, nothing like today, of course. But they had everything that hospitals had in those days. The Zinc Company built it. That's just another example of the things the Zinc Company did for the community.

Ann Trofimuk too recalls that health care was taught in school and speaks of the Zinc Company involvement with the school health program. Ann also remembers her father being hurt in the mine.

I know that it was dangerous. I think Pop was involved in a cave-in once or twice. Once he was hospitalized because he was buried up to his neck when the rocks gave way. I knew he was hospitalized, but I don't remember whether anything was broken or not. And I don't remember anybody mentioning having to pay a hospital bill, so I imagine . . . . Remember, the Franklin Hospital was operated by the Zinc Company. All the doctors were paid by the Zinc Company. Everything. They had a Zinc Company nurse, this Mrs. Hall, originally Miss Lavery, and she visited the families if they felt there was a problem. In the beginning, I think they encouraged health care in the kindergarten. I have something from our kindergarten in Franklin that I think was stamped or signed by the Zinc Company nurse. I think they made sure that children had their vaccinations or whatever. Also, remember Miss Pancoast in the Neighborhood House? Now, that was another social service that was provided by the Zinc Company.

The Zinc Company was involved with the school also. I mean, directly and indirectly. Another Zinc Company nurse was Mrs. Glynn. She was a Zinc Company nurse, but she also worked for the school. I remember if you had measles or something, she was the one in that blue uniform that came and put the sign on your door, "MEASLES," so nobody could come in or go out. See now, there was that tie-in. She worked with the school, but the Zinc Company paid her. It was a company town and the Zinc Company provided all of that.

In 1962, some eight years after the Franklin Mine had closed, the Zinc Company deeded the hospital building with its equipment and seventeen acres of grounds to the community as a gift.

He Was the Law

At that time (1915) someone was needed to bring law and order to this mining town of 4,000. And that is exactly what Herbert C. Irons did. From then on law and order reigned in Franklin with the chief keeping a firm hand on things at all times.

Franklin Borough Golden Jubilee

Mining towns have always been thought of as rough and rowdy. Just as the dangerous, dark work underground led to eerie thoughts and feelings of isolation, the transient, camp-like atmosphere of early mining towns led to boisterous, undisciplined behavior. With little other entertainment available, saloons were popular, and payday was seen as an opportunity to get drunk, and doing so often was followed by arguments and brawls that were settled with fists, clubs, knives, or guns. Boarding houses in Michigan sold liquor and encouraged miners to drink there rather than in the bars. One Calumet miner complained that sleep was impossible in the boarding houses because of the noise, commotion, and cursing "that continued throughout the night."[1]

The isolation of early mining camps meant that authority, too, was remote, or even non-existent, which allowed, or forced, men to settle disputes with fighting and violence. As small mining concerns consolidated into larger companies, however, town organization and social discipline improved. When the large corporations took command of the mines and the towns, they hired guards and detectives and established their own law and order.

Growing conflict between labor and management also led to riots and brutality. The labor struggles of the early 1900s between workers and mine owners brought about some of the most savage, bitter years in American mining history.

Other than labor disputes, however, most of the lawlessness in company towns was on a small scale. Robberies, bar room brawls, and ethnic flare-ups perpetuated a mining town's reputation for wild, disorderly conduct. Competition between miners often led to violence. Mine owners deliberately imported men from a variety of ethnic groups so as to keep them separated by language, thus preventing the formation of bonds that could lead to labor organizations. Earlier immigrants resented the newcomers and feared losing their jobs to the latest "foreigners." It was easy to blame frustrations over bad working conditions and low pay on the most recent "foreign element." Because so many miners were immigrants, ethnic prejudice was present throughout the industry. This bias can be seen in the testimonies of several of the oral history narrators as they recall the early years of disorder in Franklin. Almost invariably fights and drunkenness were blamed on "foreigners" or Mexicans.

The village of Franklin Furnace had a population of about five hundred when the New Jersey Zinc Company formed in 1897. Less than twenty years later, Franklin was incorporated into a borough with a population of about three thousand. In those years, Franklin was a typical rough mining town. Prior to the formation of the New Jersey Zinc Company, there had been little law enforcement in the town, and when incidents occurred, primarily fights between drunken miners on payday, they were handled by the Sussex County sheriff's office.

Law and order came to Franklin in 1915 with the hiring of the first chief of police, Herbert C. Irons. The prominence of Chief Irons was such that he was known not only in Franklin and Sussex County, but also throughout the state. Within a short time after Chief Irons' arrival, thirty-two lawbreakers were behind bars in the state prison at Trenton — evidence that Franklin was a rowdy, turbulent place. His illustrious career in law enforcement was so extensive that Franklin's chief of police became a legend in his own time. In spite of his toughness, however, it was emphasized over and over again that Chief Irons was good to children and was liked by them. During the heyday of the Zinc Company and the concomitant reign of Chief Irons, the children of Franklin seemed to enjoy carefree days in an atmosphere of security.

In addition to controlling crime, many of the larger mining companies supported schools, provided parks and recreational activities, and contributed to churches, all of which served as a stabilizing influence in the community.

New Jersey Zinc Company geologist John Baum recalls the influence of Chief Irons.

> Chief Irons, oh, he's a big story right there. He's worth a book. He was brought up here by the Zinc Company around the early 1900s, following the consolidation. He had experience with the Panama Constabulary down there and also with the Pennsylvania State Police. They needed someone to clean up this town because this town was a mess. This was a frontier town. This was wild. There were all kinds of fights and thefts and shootings and things like that.
>
> Many of the miners lived in boarding houses. Boarding houses were run by the widows of miners who perhaps had been killed on the job or just plain widows. And a disreputable character would come in and live in a boarding house just long enough to find out where his fellow boarders kept their funds. He'd steal it and take off. Take the stuff and decamp. Well, this was the kind of crime the town didn't like. They didn't like the brawls in the bar rooms and so on. Chief Irons was brought in here to straighten the town out and he did. He was a BIG man and he rode a horse. He was very conscious of the fact that this place was loaded with foreigners and this was America! And many of these foreigners were liable to be anarchists so he was on a search for anarchists most of the time. At one time, he was said to have gone down to the Franklin Pond and found someone who didn't speak English very well. The chief asked him if he was an anarchist, and the fellow, trying to be friendly, said, "Oh, yes, me anarchist. Me anarchist," and the chief put him in jail. This might be apocryphal; it might not be true at all. But the chief with his horse and his lasso broke up fights and was said to have lassoed a number of these people and led them off to jail. The jail in my time was in the borough hall, which was at the corner of High Street and Parker Street.
>
> The little kids liked him. The big kids, of course, sort of snarled because he was on to them. He knew what was going

on. He straightened this town out. I might tell you before I forget about it that no colored person was allowed to stay in Franklin over night. This Chief Irons, that was one of his jobs — colored folks. He couldn't get away with those methods today. But I never heard anything disrespectful in any way about Chief Irons. He was a magnificent presence.

Bill May, who clerked at the Zinc Company store for a number of years, is another admirer of Chief Irons. Bill's comments on crime in Franklin and Chief Irons clearly show his high regard for the chief.

Figure 69. Cave-ins had resulted in the collapse of the borough hall and jail in Franklin when, in the 1940s, these company houses on Nestor Street had to be demolished because the ground beneath them was collapsing. (Photo: Courtesy of Franklin Historical Society)

He Was the Law

When they brought him in here, they were looking for some tough cookie, and this guy was a military policeman in the Panama Canal Zone. In fact, Joe Herzenberg, from Herzenberg's drug store, way back I guess he was a councilman. He was instrumental in bringing Chief Irons in here because this was like a Wild West town. Fights and whatever. Stabbings. This was way back in the early years. I would say between 1910 and 1920. He used to ride around town on horseback. That was before they even had a car for him to use. He was a great big man, like a big bear. In fact, they used to call him Bearclaw. Yup. He was known all over. He was, "I am the Law," and he was too! Nobody disputed that! He was one big tough cookie.

He made a name for himself when they had the, I think it was the Silk City Gang, from Paterson, which was called Silk City years ago. That's where all the silk mills were. There was a gangster or a mobster and the state police had him surrounded in a barn in a place called "Cat's Swamp." It was around Lafayette. And Chief Irons was there. They were telling this guy, "Come on out." "We got you surrounded." Chief said, "I'm going in after him." Chief Irons goes in the barn, and they're waiting to hear gunfire and so on. Chief comes out and he's got the guy by the scruff of the neck. All the state police saw this. They all had been scared to go in there. Yet, Chief did. After that, hell, he was known all over.

In fact, do you remember "Gangbusters" years ago? It was Sunday evening [radio]. Warden Laws, who was the Warden at Sing Sing Prison, was on the show. They dramatized this particular thing with Chief Irons and the Cat Swamp Gang. That's what made him. He ended up the president of the chiefs of police of the United States.

We used to get Chief to talk about it. We'd sit in front of the theater, down here. In fact, that's one thing I really miss, the Franklin Theater. God, that was great. Anyway, we'd be sitting out front there and Chief would come around. High school kids, just sitting around. Or across the street, where there were steps going up towards High Street, concrete steps. We'd all congregate there. And that was okay as long as you weren't unruly or anything else. Chief got to know us over the years.

We always tried to stay in good with Chief Irons, because he could really help you out if you got in a scrape. We'd ask him, "Chief, tell us about the time you went in the Mexican boarding house and the guy come at you with an ax. All that stuff." And he would. Boy, could he put on a show. Boy! He says, "I shouldered in the door, and," he said, "This Mexican came at me with the ax!" He said, "If I stepped back, he'd a cut me in half, but," he says, "I went in with a right hand and broke three ribs, and I proceeded to mop the place up with him." We'd be there trying to keep from laughing at him because he was so serious about it. But he was a good guy to know, I'll tell you. Boy, he kept a lot of us out of trouble because we knew, "Don't deal with Chief Irons." He would just as soon put you in jail as not. Or scare the hell right out of you, you know, when you're a kid.

Chief Irons, he was like two guys today. In those days, everything came in by train. There used to be a big train station, the Susquehanna station. But that's burned down. That's all gone. Yup. Oh, it was a big place. You could play ball in there. [The Stationmaster] didn't care as long as we didn't bother anything. He had his own housing right in there. There was the apartment. There was the office. There was a waiting room. A waiting room, can you imagine that? For people to sit and wait for the trains. That's what a big deal it was at that time.

Then there was a big freight house on a [railroad] siding. I can remember they were bringing in cars on the railroad cars for Lorenzos. Bringing in Oldsmobiles from Detroit. Today, everything is in trucks. But in those days, everything was in big freight trains. In fact, the Zinc Company payroll would come in by the train. Chief Irons would be there with his buddy — the other cop was Ernie Duck — and they'd bring out the strong box and the big bags of change and all that. Put it in the police car and take that to the bank.

Tom Sliker, the last mine superintendent at Sterling Hill, also remembers that Chief Irons would meet the trains in Franklin.

When I came into Franklin, the first time, you were met by the chief of police here, which is Irons. If you didn't have no business that he thought was business here, you didn't stay here. You got out of Franklin. He didn't let nobody stop here. They used to have a depot down here and trains used to pull in and out of Franklin. He was there when the trains come in and anybody got off that train that he didn't know, he had to

Figure 70. Chief Irons would meet the trains and make certain that no undesireables got off and remained in Franklin. (Photo: *Zinc Magazine;* Courtesy of Ogdensburg Historical Society)

know what their business was, or else they got on the train to move on.

In fact, he was brought here because back in those days — that's before my time here — the Mexicans had a lot of little problems of their own, I guess. So he came to straighten that out. He used to ride horseback at that time. He done a good job.

He was a good friend of mine, Mr. Irons was. During the war, the war came on as soon as I got in the mine and I was deferred by the company. I had a 1936 car and I was going to sell it. So I said to Mr. Irons, "Do I have a right to sell my car over price?" He said "No, you're not supposed to," he said, "But why are you selling it?" I said, "Well, I can get pretty good money for it." "Tom," he said, "If you get more money for your car and you have to go and buy another car, you're going to have to pay the same as that guy had to pay for your car, so," he says, "If I was you, I wouldn't do it." So he was a friend of mine for the rest of his life. A very nice man. Rough man, but he was good.

Eddie Mindlin also recalls the legendary Chief Irons, and lists some personal incidents involving the chief's law enforcement. Furthermore, Eddie verifies the story that Chief Irons met the trains coming into Franklin to assure that only people that he thought were acceptable remained and to make certain that no African Americans got off. As Eddie explains:

> If they were colored, that was a taboo. The only one that walked through Franklin, and he was only to go through Franklin to go to the church, was George Piggery. Do you remember George Piggery? A nice black man, a gentleman. He lived in the Monroe area and he used to walk the railroad tracks and come up and go to the Catholic Church. Couldn't stop him. But there was never a colored person in town while Irons lived here. There was in Sussex. There was in Newton. But, in Franklin, he was the law.
>
> My father told me, in the early days, "Nobody walked the streets at night." It was like a Wild West city because there was drinking, mugging, stuff like that. The miners were rough. They imported laborers. That's where a lot of the

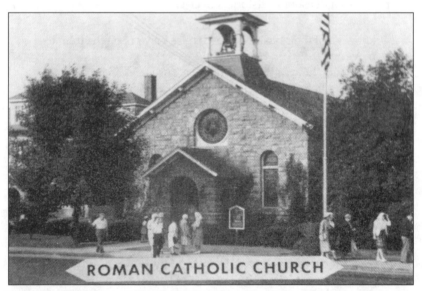

Figure 71. The Roman Catholic Church in Franklin. The only black person who Chief Irons allowed to walk through Franklin was George Piggery, a parishoner of this church. (Photo: *Zinc Magazine*; Courtesy of Ogdensburg Historical Society)

knifing came from until he came into town. They imported Chief Irons. He stopped all that. Pop said he used to ride a white horse. He could talk to someone by pulling a gun on them. Someone was needed to bring law and order in the mining town. He was a policeman on the police force at the Panama Canal. There was no law. He made the law.

In Prohibition times, everybody made booze. Everybody. My father liked his own schnapps. One customer came in and he was drunk. He wanted a drink with Pop. Customers that came in, drinks were free. But he was drunk and my father wouldn't give him a drink. He says, "Sam, I'll squeal on you. You make your own." Pop says, "You're not getting anymore, you're drunk." So, somehow or other, he told Irons and Irons came because it was a complaint. Irons came in and he found paraphernalia and he put my father in jail overnight. Prohibition.

Irons, he made one law, NO RIDING DOUBLE on a bike. Coming through the park from the school, as soon as I got on High Street, he saw me come up, and he caught me [riding with another kid on the bicycle] right there, at the top of High Street and Junction Street. Took my bike away. My father had to go get it.

But he was a knockover for kids. Really, he was. He loved kids. If it was late at night around the Neighborhood House, he'd say, "It's time to go home." No hanky panky with Irons.

As a child, Steve Revay also had a few run-ins with the chief and tells about several incidents.

I was born 1907. I was ten years old so it must have been about 1917 or somewhere in there. At the vocational school, I always used to hang around the ball field. I liked ball players so I hung around there quite a lot and everybody knew me because I was always around. One summer night in the fall, I was at home and I was sawing wood with my father to prepare for the winter. In them days, you couldn't buy wood that was sawed. You'd have to saw it yourself and split it up for winter use. Well, we were there sawing away. This was on Rutherford Avenue. I could look up and see the grandstand up on the ball field. I looked up there and first thing you know, I saw smoke piling out of the grandstand. I said to my father, "Look at that, there's smoke coming out of there." He says, "Yeah, it looks like a fire." So we kept sawing.

Then he says, "Aaah, get out of here," he says, "You're jerking the saw anyway. Go on up." So I went up there and hung around and the grandstand was on fire. They come up there with the hose and they're pulling this cart. But they didn't have no water. The grandstand burned right down to the ground. This was on a Friday night.

Well, Saturday morning, I was sitting on a bench outside, and first thing you know, Irons and Straulina — Straulina was the Zinc Company spy — drove up, and Irons says to me, "Come here!" So I went over and he grabbed me, opened the car door, and threw me right in the car. So I'm wondering, "What the hell is all this about?" Then he started on me. He says, "You set that grandstand on fire, didn't you?" "You burnt that grandstand down." "No," I said, "I was home sawing wood with my father when the fire started." And he kept hounding me and hounding me.

Then he took me up to the borough hall. He was a burly man. He was a mean man, you know. He was a good cop, but he was mean. I've seen him beat people up. Well, anyway, he took me in the borough hall and he questioned me about the fire. After he got through with me, he grabbed me and he took me in the jailhouse. They had four or five cells in there. And he opened one of the cells, pushed me in there, locked the cell, and went out. Here, a ten-year-old kid, he locks him up in jail and the kid don't even know nothing about it. That's how mean he was. So I was in jail there till my mother come up about three or four hours later and then he let me go. And they never did find out who burnt the grandstand down. That was my first experience with Irons.

Then I had another experience with Irons. He was always chasing me for some reason or other. He must have thought I was a bad guy. There was a boxing match up in the garage there [Lorenzo's Garage on Main Street]. Upstairs. They had a place where they used to hold boxing matches. Well, anyway, one day they had a boxing match there. I was a young kid, you know, and I wanted to go see it. I didn't have no money. I couldn't get in. So I went up the fire escape and I was peeking through the window to see what was going on. Then, first thing you know, there was two policemen there. Pete Lanaman, he spotted us up there. He said, "You kids get out of here and get home." So we come on down. And when we got in front of the garage on Main Street, who was there but Irons. He said, "You kids get out of here. Get home. Go on! Get going!" He expected us to run, you know.

But I didn't run, just was walking and I was walking easy. And first thing you know, his two hands were on my shoulders and he was pushing me. And he was really pushing me. By him pushing me, we were going like the devil. And he stumbled on something and he fell. I could hear him go KERPLUNK, and boy did I leg it then. I run like hell and come on home. But, he was chasing kids all the time and I was always afraid of him after that. Whenever I saw him, I would run away from him.

When he first came here, he used to ride around on a horse. Well, first he walked. Then he got a horse. And then he got a motorcycle. Well, here on this corner, [Main Street and Rutherford Avenue] I was standing there one day and he came down the road with the motorcycle. When he went to turn, to make the turn to come up Main Street, the motorcycle slid from under him and he went a- - over head. I was standing on the sidewalk and I couldn't help but laugh, and I laughed. First thing you know, he got up and he started for me. And he chased me halfway up the street because I laughed I suppose. I got away from him.

When the council got Irons, he came out of the army. They said he was in Panama. They wanted a tough cop and they got a tough cop. They said his name was Irons, but I still believe it was some other name. I think he's some kind of a Pollock. My mother wouldn't take no baloney from anybody. She had a neighbor who had a boy and there was another neighbor across the street who had a boy and these two boys got in a fight. Well, the boy that got in the fight was an English boy and this other boy that my mother was sticking up for was a Slovak. And Irons had this [Slovak] kid and he was trying to take him away, and my mother kept telling this woman in Slovak, "Don't let him have the kid." "You hang on to the kid." "Don't give the kid to him because he'll take him away and put him in jail." She kept telling this woman in Slovak, "Don't let him have him!" "Don't let him have him!" And that damn Irons, first thing you know, he turned on my mother and I thought he was going to hit her or whip her or something. I think he understood every word she said. So that's why I say that his name wasn't Irons. Maybe it was Ironski.

During Prohibition — you know how the miners are — they're heavy drinking people. So when they passed the law [Eighteenth Amendment in 1920] and Prohibition come into effect, they didn't have no where to get their booze. So they started to make their own. They had a guy in Franklin, he was a plumber, he used to make stills. He'd make these little stills for them and they'd make their own booze. So Irons started to raid these people. He'd raid them and take them to jail and fine them and so on.

Of course Irons, he was the policeman, he was the jury, he was the whole damn works. You didn't have no jury or nothing. If he said you were guilty, you were guilty, that's all. Anyway, he began to raid them and he'd take them over to Newton [the county seat] to fine them. And they'd take a lot of money out of here. So we had a mayor, his name was Jenkins. The newspaper come out with a Newton budget for the fiscal year and on that budget, they had estimated income from Franklin, for fines for having stills. And it was a lot of money. A lot of money and that money was all going out of Franklin, going to Newton. So the mayor, he got peeved off about it and he gives Irons hell. He says, "No more raiding stills," he says, "Because all this money," he says, "is going over to Newton, and now they've even got an estimated income coming to them from Franklin." So they cut it out.

Bob Shelton, the retired Franklin banker, presents some of the chief's good and bad points. He too had some personal dealings with Chief Irons. Bob recalls:

He didn't talk to me for two years. A young lawyer got in trouble with him and he bossed the young lawyer around. What happened was he was defending someone. Anyway, I was going to testify [against] what Irons said — I don't even remember what the hell it was. Well, he never forgave me.

He was a rough customer, but he never *buffaloed* me. I told him, I says, "You just draw that gun once, boy, and you'll find yourself in the electric chair. Don't think you can *buffalo* me. I know where you come from, the old mill, Brooklyn." He was a Brooklynite. He didn't know where I was from. I said, "What the hell do you think you are running East New York again?" But he was a good cop.

We needed a tough cop when he came. He roughed up these people very badly. With the miners who were rough and sometimes beat up their wives and things like that, he came in and he straightened them out. We had two taverns and they would drink there. Sometimes they would get drunk,

Figure 72. Narrator Robert Shelton and Mrs. Shelton. Chief Irons, Bob says, was "very good with that teenage crowd." (Photo: Carrie Papa)

but we never really had any bad trouble. Irons was — well, he's no different from a banker. Sometimes you can get ugly and sometimes you're normal. When he was normal, he was a very pleasant man. When he first came here, this was a tough town to handle because it had had no discipline. He did control that.

He was very good to the kids. One time, my boy was in trouble. One time, the kids had firecrackers. It was against the law. Still is, I guess. But the kids went up to New York State and got firecrackers. Well, when Irons found out that Bob had firecrackers, he came down to the house. It was so funny. The kids were so frightened. Oh God, they were scared to death. But, he came in and talked to them, why we needed rules and what damage a firecracker could do. You know, your eyes or your fingers. And this is the way Irons was. He was very good with children. Very good with that teenage crowd.

Harry Romaine relates a short story that reveals the more charitable and human side of Chief Irons. Furthermore, this incident shows the close connection between the Zinc Company and town officials.

My wife Joyce and I come down Main Street in Franklin one time and Joyce says, "We need a loaf of bread." So I said, "All right, I'll stop in front of the Zinc Company. You run in and get a loaf of bread." So Joyce went in and got the bread. Just about this time Chief Irons come from High Street down and up this way. He stopped right in the middle of the road. He bawled me out! Up hill and down! For parking in the Zinc Company parking lot on Main Street. Well, anyhow, he just bawled me out. No ticket or nothing.

I lived over on what they called Scott Road. We went home and I pulled up on the side of the road to let Joyce out. Who was right behind me but Chief Irons. He come and apologized. He says, "You know, that's my bread and butter. The Zinc Company." The Zinc Company, they didn't hire Irons, but they controlled everything. All of the bosses, they managed to get on the council as mayor and councilmen, you know. It was a company town.

Another view of the personal and unexpected side of Chief Irons comes from Alan Tillison.

I talked to him several times and I always thought he was a reasonable person. One incident I do remember before I got to know him. It was when B. D. Simmons had the place right there across from the post office. Well, on the side where the post office was, they had signs, NO PARKING. So I pulled up and going from this direction, I pulled across and parked by B. D. Simmons. I wanted to go to Simmons. About that time, Irons pulled up right in front of me. He got out and he says, "Hey, you know you're parked the wrong way across the street?" And I said, "Well, Mr. Irons," I said, "I didn't figure I could park on that side." I says, "It says NO PARKING by the post office." "Well," he says, "We've got signs NO PARKING there," he says, "But they do park there and we don't bother them none."

Then he went over to the post office and I went in Simmons. And just as I came out of Simmons, he was coming down the steps from the post office. "Hey, wait a minute," he says. And I thought, "Oh, oh, something's going to happen." He come over and he says, "You know, I've been thinking about it and," he says, "I think you were right."

In the beginning, he had to be very rough, I think. See, originally, when the mine started they brought in a lot of

Mexicans and they had a lot of trouble with them. He was brought in because they did have problems. But he shaped the town up. That's for sure. [With Irons here] we didn't worry about locking our doors.

Perhaps due to the efforts of Chief Irons, there was little fear of crime in Franklin and people did not even lock their doors. Steve Novak's comments are typical.

> One thing, back in the old days, you never had to worry about crime. Never had to lock the doors on your home. Everybody helped one another. If anybody needed any help, just say something and pretty soon you had a gang wanted to give you a hand. Everybody was considerate with one another. In fact, a friend like Doug Geary or Charlie Morris or Gil Snyder come to the house and nobody home. Door's unlocked. They'd make themselves a sandwich and leave a note. "We were here and had this and that." Thought nothing of it.

The words of Laura Falcone could have described several other mining towns as well as Franklin. The sense of security that Laura felt was confirmed by numerous mining families.

> We were never afraid. One thing we grew up with was lack of fear. When I went away to school, I was amazed that everyone locked their door. Our house was never locked. And not only ours, but I would guess 95 percent of the homes in Franklin were never locked. We went all over — Mom allowed us to go all over — without ever being afraid. We would spend all day over at the park, walk down to the pond, go for hikes in the woods, pick violets at the "Wildcat," and go anyplace in town, without Mom knowing just where we were at any given time.
>
> As little kids, we often went together and Mom would make us a lunch to take with us. Mom made us take Evie too, in the baby carriage, so she must have been really small. I remember lugging this baby carriage up behind the bank on the way to the pond. These were not roads, but country trails. This was before Buckwheat Road was put in. After going through the area behind the bank, we would come out around Evans Street, where the mineral museum is today,

Figure 73. Chief Irons leading the Halloween parade. According to narrator Laura Falcone, Chief Irons "was a friend of all kids." (Photo: Courtesy of Franklin Historical Society)

cut through Katzenstein's fields and arrive down at the pond. This would have to be at least a couple of miles each way, and quite isolated.

I don't suppose we wandered very far from home at night, but I do recall playing outside after dark. We would play hide and seek, or kick the can, or sardines. All the kids in the neighborhood and all over the neighborhood. But mostly we were in the park. That was always attended so it wasn't just an empty park. There was someone there, a counselor that you could go ask for games and toys to play with. There was always a policeman around. Mr. Irons, I believe his name was. He would push us on the swings. He was a friend of all the kids. We liked him too and we were proud of him.

Whether people liked or disliked Franklin's chief of police seemed to depend on their particular dealings with him. But, whatever one's personal opinion, no one disputed that he had a powerful influence on the town and everyone credited Chief Irons with bringing law and order to Franklin.

1. Thurner, *Calumet Copper and People*, 59.

School, Sports, and Personal Goals

There's a high school on the hilltop
  Where the oak tree stands so strong
  To her prestige and her glory
  Raise we now our voice in song.
Franklin High, our Alma Mater
  To thy name we'll e'er be true
  Loyally we'll guard thy honor
  And our colors, white and blue.
When we leave these halls we cherish
  And our paths lead far apart,
  Still the memories of our high school
  Will remain within our heart.

Franklin High School Alma Mater

Mining enterprises, as the largest taxpayers in the company towns, had vested interests in the public schools, and they oversaw this interest by having company officials serve on school boards. This certainly was true in Franklin where mine superintendent Clarence M. Haight also was chairman of the board of education from 1941 to 1949. Several other high officials with the Zinc Company also served on the school board. This presence gave the Zinc Company enough influence to see that a vocational school was established in Franklin at a time that very few small mining communities had high schools of any kind. For the most part miners themselves were not educated and did not expect their children to have much schooling. As late as 1947 the children of many mining families, particularly in Appalachia, were still attending one-room schools.[1]

Figure 74. Franklin High School in 1950. (Photo: *Zinc Magazine;* Courtesy of Ogdensburg Historical Society)

By establishing vocational schools, the mining companies were taking steps to insure that the children of miners would be prepared for the type of jobs that were, or that would be, available in the area. While it may have been self-serving on the part of the mining companies to advocate these schools, they did provide educational opportunities that otherwise would not have been available in these communities. The Zinc Company, although unusually progressive for the times, was not alone in supporting a local school. Even in Appalachia, a few mining companies like the Pulaski Anthracite Coal Company in Virginia were building public schools. In the western states, many large mining corporations constructed dozens of schools. The Union Pacific Coal Company erected the first school building in Hanna, Wyoming. In Michigan, the Calumet & Hecla Mining Company had built twelve schoolhouses by 1914. Although they originally were "Manual Training Schools," by the 1950s the Calumet & Hecla schools were offering industrial, commercial, and college preparatory courses. The school system eventually became "the envy of many other places of greater pretensions and much greater population."[2] A number of narrators displayed this same kind of pride in believing the Franklin Vocational School to be the envy of other places. Contrasting this view, a few narrators felt that the "vocational" school was of more benefit to the Zinc Company than to the people.

Prior to the establishment of Franklin as an independent borough, children had been sent to a two-room school known as the Oak Street School. During the early years of the Zinc Company, a kindergarten was housed in the Neighborhood House. The Franklin Vocational School opened in 1915 and included an elementary school. The new vocational school contained machine and wood shops for the boys and cooking and sewing facilities for the girls, as well as standard classrooms. Music and art departments, a library, gymnasium, and an auditorium equipped with a "moving picture machine" made the school the most modern and best-equipped in Sussex County.

In 1922, a junior high school, which included the ninth grade, was added and, five years later, a senior high school was constructed. Prior to that time, in order to attend a high school it was necessary for students to travel to other towns. In 1927, seniors studying at neighboring high schools were brought back to Franklin to complete their high school education and become, as the class of 1927, the first senior class to graduate from Franklin High School.

Steve Revay attended the small Oak Street School before the vocational school was built. However, Steve did feel that building a vocational school in Franklin instead of a conventional high school was not the best thing for the community. Steve observes:

> I went to the old school down by Littells. I went there for about a year and a half. Then they built the school up here on the hill — the vocational school. When I went to school there, I was about ten years old. They called it a vocational school and they had vocational training like carpenters and machinists and all that stuff. That was intended to help the Zinc Company produce labor for the mine. Instead of a high school, they wanted a vocational school because they wanted to get people to learn a trade so they could work for them. That's why they didn't want a high school. I think they made a big mistake when they built a vocational school.

Figure 75. Narrator Steve Revay and Mrs. Revay. Steve felt that the Franklin Vocational School was intended to help "the Zinc Company produce labor for the mine." (Photo: Carrie Papa)

If you wanted to go to high school, you had to go to Newton on the train everyday. I went to Newton High School. Then I had typhoid fever. When I got well, I would have missed the whole year of school so I thought I'd go work for awhile and then go back to school afterwards. But I never went back to school because my mother and father needed the money. I was fifteen years old when I started to work for the Zinc Company.

Later on, they found out that [the vocational school] didn't work because nobody went to the vocational classes. They all wanted to go to Newton to the high school. So they changed it [the vocational school] to high school.

Bill Dolan, on the other hand, benefited from the vocational school system. Bill was born and raised in Ogdensburg and attended the grammar school there. He started working at Sterling Hill when he was barely sixteen years old. As an adult he was given the opportunity to obtain further education at the Franklin Vocational School. As Bill explains:

I went to work for the Zinc Company when I was sixteen. Later, when I was about twenty or twenty-one, the Zinc Company gave us the opportunity to go to the Franklin

Vocational School to take up machinist, carpentry, brass-cutter, whatever, so I went to the machinist shop. For two years. I would say about 1923, maybe. The Zinc Company paid for us to go there and they paid for the class. We didn't get any extra money from work. They just gave us the opportunity to go there and learn whatever we wanted to learn. I was employed during the day and then we went there at night.

One of the teachers there was my boss in later years. Ben Penberthy. He was teaching night school, but he was still employed by the Zinc Company. One of the men running the machines had a heart attack and they gave me the opportunity to go down there [Franklin Vocational School] and learn to do the work. This man Penberthy was an exceptional boss. He learned his trade in England. Then he came to the United States and he was boss in Franklin for a long time.

Sterling Hill miner Tom Sliker studied with Ben Penberthy at the Franklin Vocational School. Although Tom originally wanted to be a machinist, he was unable to complete the necessary training. Instead, Tom advanced from being a miner to become mine superintendent and, as a result, was able to afford a better education for his children than was available to him. As Tom reports:

> When I first got married, I started going to night school. I wanted to be a machinist. That was the only thing I ever wanted to be was a machinist. Before I went to the mine, I was going to the school in Franklin, night school. I was working on a farm. I used to get up three-thirty in the morning, and I couldn't keep it up because it was too early. Working all day and then come home and go to school till ten o'clock at night. Penberthy was the lathe operator and the shop man and he was teaching the classes at that time. I had good classes while I was there, but I couldn't keep it up.
>
> When I went there [Sterling Hill], I loved it, and I learned everything that you could learn about mining. Ted Holmes, was a superintendent in the mine for probably eight, ten years. I was talking to him one day, "Well, I'd like to have had a college career," and he says, "You had one." He said, "You might not be able to have it on paper, but you have it in

your mind because you're one of the best miners I've met in my life."

My daughter and son both went to Franklin School. My daughter went to nursing school in Montclair, and my son went to college at Susquehanna University in Pennsylvania. In our street here, I suppose we must have had twenty young men and women, and I think it was just my son out of all those twenty that went to college.

After serving in the navy during World War II, Eddie Mindlin attended Franklin High School. He remembers the early two-room school and provides some little known facts about the original Oak Street School. Eddie also recalls the use of nick-names in high school.

> That building that was next to the synagogue on Oak Street, that was the first school. Then, after the school [closed], it became a silk mill. Then it was torn down. Joe Kulsar got the bricks from that and built his gas station with those bricks.
>
> I quit [Franklin High School] and joined the navy. I was there three years. I came back out and I graduated in 1947. We had a group that was called The Vets in high school. Franklin was really something about nicknames. Everybody

Figure 76. Narrator Eddie Mindlin and Mrs. Mindlin. After three years in the Navy, Eddie came back to Franklin, attended Franklin High School, and graduated in 1947. (Photo: Carrie Papa)

had a nickname. Like Tetertote. I forget his first name already. Francis Fasolo. His brother Tommy was in my class. Tilly Fasolo. How he got Tilly, I don't know. There was Alec Benzowich. He was called Shegator. What's a shegator? A guy that goes after the shes is a she-go-getter. There was a guy called Popeye. There was Moon Kinney. There was Harry the Hat. Sashey Stankowich, Sashey wasn't his real name. Sashey was a nickname because it was a Russian name. There was so many of them. Poo Shelton. Pete Nestor, his name was Ed Nestor. Ninety-five percent of the people had nicknames.

I went to Washington with the 1945 class. Florence Mitchell, Flo, always called me Edso. And talk about nicknames, there was a group that called me Jewboy. Richard Hendershot used to say, "Hey, what's new, Jew?" and I used to say, "Nothing new, Ninety-two," because "92" was his football jersey. It didn't bother me. It was just a game.

I've always bragged about our high school, which was because of the Zinc Company too because they ran the board of education. They ran everything. You have to understand that the Zinc Company kept Franklin alive.

Former Zinc Company store employee, Bill May, remembers nicknames and sports as being a distinctive part of his Franklin High School experience.

Years ago, it seemed like all the boys had nicknames. Girls didn't have them that I know of. One for instance would be Frank Nemeth. His name everybody knew was Hotdog. Hotdog Nemeth. A kid got killed on a bicycle, and I heard "Frank Nemeth got killed." It was like I never heard the name before. "Yeah, you know him, Hotdog." Oh, God, because I knew this kid. I caddied with him at the golf course. Everybody had a nickname. We had a janitor in the school, WaWa. Joe Garnes, WaWa. Who gave him that name? This guy's name, Shegator. Where'd you get a name like that? I don't know. He's got a brother, Pockwah. Where'd these names come from?

I was only going through high school when the war was on, the Second World War. I graduated in 1946. I wanted to get the GI Bill of Rights and go to college. That was the idea. But I had my one bad left eye. I talked two fellows into going down with me to sign up. They passed and I didn't. I said to the guys, [recruitment officers] "Bad eye," I said, "During the war, you took fellows that were blind in one eye and couldn't see out of the other." He said, "Well, this is the way the army is now, more strict." Not getting the GI Bill through the service, that put it [further education] out of sight for me. I said, "Oh, boy, what am I going to do?" So I ended up getting a job in the Zinc Company store. I had a couple of friends that were in the store and they started that way too.

They had a miner's football team that put Franklin on the map, I would say. Sports-wise, anyway. I played six years with the Miners when I got out of high school. *We're a bunch of miners, miners are we. We hail from Franklin, Franklin University.*

Practically every little town in Sussex County had a baseball team. I can remember up at the field here, the old high school, Franklin High School, they had a big grandstand. It could seat about six hundred people. That's long gone. Franklin's always been a hell of a sports town. My brother Bob got a football scholarship to the University of Maryland. A lot of good ball players come out of here. They held tryouts for the Philadelphia Phillies. Had a tryout at the Franklin School for baseball. Scouts would come up and see who's who and what's what. They had some sports teams over the years. Billy Glynn played for the Cleveland Indians. Russ VanAtter played for the New York Yankees. Babe Ruth used to come up here to the Wallkill Country Club and play golf. In fact, I've got his autograph on a Wallkill Country Club score card. I caddied there. I started there when I was about ten years old till I was about sixteen.

Steve "Itchy" Novak was one of the original members of the Franklin Miners football team. Steve wistfully recalls being denied an opportunity to accept an athletic scholarship that he had been offered, and remembers that Babe Ruth often came to Franklin. As Steve recounts:

> I had a chance to go to Montclair State. We had a small track team. It was a club team. So we had a track meet with Montclair. It was a practice meet for them, and the coach was impressed with my ability. Wanted me to go to school, but I didn't have the credits to go. In the early days, there wasn't much guidance. Everybody was on his own. Then when I

Figure 77. Narrator Steven Novak was one of the original members of the Franklin Miners football team. (Photo: Carrie Papa)

got out [in 1930] I had a chance to go to school on a scholarship, but I didn't have enough credits. I went to the principal and asked him if I could take a post graduate course and take up some language and some algebra, and he wouldn't let me. So I had to go and get a job and go to work.

About the founding of the Franklin Miners, one day after mass, Lou Riggio — Lord, have mercy — he's dead now. He was killed in the plant with the Zinc Company. He said, "We have a football game this afternoon." I said, "A football game, we don't even have a team." Well, he had been down to Wharton talking to fellows and they were telling him what a great team they had and he said, "We have a team in Franklin to beat you guys!" Well, when Lou came to me that Sunday morning, we didn't have any team, so we rounded up a bunch of fellows that Sunday morning. Went up and practiced for about an hour or so with whatever we could get together for uniforms and protective clothing. Went down there and played and we beat them. Then Andy Fedor — he worked for the Zinc Company — took over coaching the team. So we got a pretty good team. Called the Franklin Miners. It became the town football team. That was the first Miners team. They had a pretty good following. We must have played ten or fifteen years.

They had the Franklin Miners Band, which was a local band. They used to get dressed up in miner's clothes, and miner's hats and they played college songs at the football games. I remember one time the Dover Field Club was going to play Irvington Camptown down at Irvington. Dover had a good fife and drum corps, a good band. But the Dover Club asked the Franklin Miners Band to come and play for them, because they were that good. Lou Riggio — the Lord have mercy on him — he was the leader. He had a baton made out of wood, like a miner's pick, and he used that as the band leader. All of the fellows were dressed up in miner's togs and they paraded around the field just like in college. I was talking to a policeman. He says, "This is the best damn show we ever had here."

One of the reasons Babe Ruth liked to come to Franklin and Sussex County was because he could be by himself. [Looking at photographs] That's Babe and Russ VanAtter. He was from Franklin and he played with the Yankees. He was a pitcher 'til he injured his hand. He had a place at Beaver Lake. There was a fire and he went to save his dog and he cut his pitching hand. That ruined his career. Here's another picture of Babe. He was a good hunter. Good shot with a gun. He shot a pheasant and is reaching down to get the bird from the dog. They played a trick on Babe that day. The same day that he shot the pheasant. I went around with Babe and they were supposed to bring him around in a circle. It was November and the bog was frozen. This was on Cal Hagart's farm in Hainesville. Then they were supposed to bring him to a clearing where there was a stone wall. Behind the stone wall, they had this turkey. They planted it there. All you could see above the stone wall was the head of the turkey. Doc Huff says, "Babe, there's a turkey over there." So Babe got it the first shot. When they got back, they weighed the turkey. It weighed thirty-four pounds. And he never knew it [was a joke]. He kept saying, "Damn it, I've hunted before, but that's the dumbest damn turkey I ever saw." He says, "I'll bet the damn thing wasn't wild." "Why sure, it was wild!" "Why did it stand there and look at me like that?" "It probably never saw anybody before." Well, he never knew it until Bob Edge, a sports commentator on WOR, spilled the beans on WOR one night. And that made Babe so damn mad, he said, "Why did he have to tell the whole world that they made a monkey out of me?" He [Babe Ruth] met so many fellows here. He used to bowl here too. This picture is at Russ VanAtter's place at Beaver Lake.

John Baum corroborates the influence of the Zinc Company in having a vocational school established in Franklin. John further notes how some students were able to go on to college on sports scholarships.

> The school, of course, was established by the company. As was the hospital, and so on, the whole town. This was a vocational school, theoretically. It was built by the mining company to teach the children a vocation, be it carpentry or another trade. The community needed carpenters, plumbers, and things. In many cases, people learned their trades at the Zinc Company. Until very recently, if I needed a painter, I'd go hire a retired Zinc Company fellow who had been in the real estate gang. Like everybody in the Zinc Company, they knew everything, which goes to say anybody who was in the real estate gang was not only a carpenter. He could paint. He could plaster. He could do anything.
>
> There was no thought that these people [miners' children] were going on to college, but a lot of them did. The big hope for many of them was football scholarships. We had, as I say, Hungarians here and various eastern European groups, and these people were ideally built for football. A number of these people went on to college on football scholarships. One of our most prominent lawyers here, Don Kovach, Donnie went to the University of Virginia on, I believe it was, a football scholarship. And a number of others did too. Each generation has improved itself.

Don Kovach, former miner and lawyer, and current CEO of the National Bank of Sussex County, is one of those who attended college on a football scholarship. While Don attended the vocational school after it had been accredited as a high school by the State of New Jersey, he defends the Zinc Company's development of a vocational school in Franklin. Don credits his success in life to the teachers and coaches who encouraged him to make the most of his athletic ability.

The first school system in Franklin was a vocational school system and it was dedicated exclusively to training the young people of Franklin, which wasn't such a bad idea at the time. Remember, there wasn't a whole lot of opportunity. To be trained as a draftsman or lathe operator or other type of machinist meant that you could be productive. From the administrative end, the women being trained in commercial typing, bookkeeping, and so forth all worked to the benefit, in my opinion, of a one-industry town. Many of those whom I knew that went through the vocational system or later on through the high school system, did so, in fact, with a great deal of success. Some moved up through the ranks — working for the Zinc Company.

That's the way the town was structured including the educational system. It was a very parochial, paternal type of environment. The school slanted you. First of all, there was a strong work ethnic. You knew you were supposed to work. That was part of your obligation in life, you worked. And secondly, the idea that you had to be better — maybe because of this, let's call it, latent discrimination — I think that made you more competitive. At least I felt that way. And I was told by my family, "Look, you know, you're going to have to do

Figure 78. Narrator Donald Kovach explains the subtle discrimination that was built into the Franklin school system. (Photo: Carrie Papa)

better than some of the other people around you because those that make the decisions as to what happens to you don't necessarily favor . . . ." Anyway, there was a discrimination that ran throughout the community.

For example, when you started school, there was an "A," "B," and "C" class. Invariably, those that were in the "A" class, which was considered to be the advanced class, were the bankers', doctors', merchants' children. "B" class would maybe be the people who were more blue collar workers, maybe in the Zinc Company administration end of it, but not necessarily on a high intellectual level. And then there was the "C" class. Most of the "C" class was the miners' children.

However, as you progressed through the school system — now I don't know whether it was because the times were changing or the educational system was such — but, if it was determined that you had capabilities, aptitude or IQ, you could move through the ranks and end up in the "A" class. It took me until seventh grade. In seventh grade, I was promoted from Six "C" to Seven "A."

And I found a whole new level of friends, people I had gone to school with but never associated with. I mean, all of my friends, I think almost exclusively up until that time, were miners. Then, when I got promoted to the "A" class, I met a different group. I knew they existed, but I never ran with them, if you will. I can remember it was daunting to go into a [different] classroom in seventh grade, where I always felt that the group that was in that class were sissies and not part of my group. I must have been thirteen, maybe. I never even considered that they might be somewhat intellectually advanced because of their environment. But, in any event, that [that they were sissies] turned out to be a false assessment because some of my best friends are still those that I went through high school with. It became, in high school, more homogeneous.

Seventh and eighth grade was a precursor to ninth grade. They [the teachers or administrators] kind of determined where they thought you ought to be. So, when I went in Seven "A," I guess my progress was such that when I entered ninth grade I was in the college preparatory section. I think athletics had a lot to do with me reaching a point where I felt I was equal, or in some instances, maybe even superior to my peers. Previously I thought otherwise. So then I felt, "Gee, if they can do it, I can do it," that kind of thing.

It wasn't until probably my junior and senior years in high school that I realized I could go to college. I also knew that if I wanted to go to college, I better find a way to finance it. I had help from some of the teachers. I could name my football coach who was John Kolibas. He was a tough, I guess Slovak, from Carteret who coached us. Our superintendent was Mr. Hollobaugh. Our baseball and basketball coach was Reginal Purdy. We had other teachers who had connections, like John Bennett, who became superintendent of the schools in Toms River. Another influential teacher was Clarence McKeeby. All those gentlemen had, at one time or another, been involved in coaching and athletics. Also, they were very interested in seeing young men and women who had talent go on. They would help guide us.

One of the rich parts of the sporting history of the community — and it has indeed a very unique history — is that several Franklin High School or Franklin residents played major league baseball. One thing that is still prominent — if you ask anybody fifty years old or above, or perhaps even younger — they will tell you that they remember the Franklin Miners. The Franklin Miners were just what that connotes. They were a team that was generally fellows that would come up out of the mines and practice. They would challenge anybody. Their reputation became statewide and beyond.

Now, what I think is that the high school team tried to emulate the type of football that the Miners played. That was just tough, hard, what they call smash-face football. During the period of time that I was in high school, we had a decent team. We had some big kids. One went to Blair Academy and then went on to the University of Virginia. Our team had three players that were awarded full scholarships related to their athletic ability. One went to the University of Delaware. That was the fellow that I said became a brigadier general, Bob Gunderman, who was a classmate of mine. And I went to the University of Virginia on a scholarship.

I think it was because the kids were hard-nosed kids and I also think it had something to do with the academic training that we had. Although the University of Virginia was a receiving college for prep school students, particularly the elite of the South and the first families of Virginia so to speak, I had no problem competing academically. You did not go there and get by simply because you were an athlete.

We [in Franklin] had a good school system. There were sixty-five or seventy people who graduated with the class of 1953. In that class, one of the members became a brigadier general. One of them is still a state senator. I became a lawyer and now I'm CEO of this bank. One went to the University of Michigan and became a lawyer for a large utility in the state of Washington. Another served as a career officer and retired as a bird colonel. One became a chief of police of the town of Franklin.

State senator Robert E. Littell agrees with Don that the establishment of a vocational school was warranted in Franklin. Bob defends the vocational school and adds that it helped him become successful in both high school and the military.

That [vocational training] was good. I took college prep courses when I went to high school, but you also had to take metal shop and wood shop. The advantage of that was that it taught me hand-eye coordination. It taught me how something is made from a piece of raw material and turned into a finished product, like a lamp, or a dresser, or a box, or whatever it was. Those are important skills that serve me well today.

I have very fond memories of my time in the Franklin public schools. I think a lot of the young people who graduated from Franklin High School went on to college. During the period of time when I grew up, many went into military service because of the Second World War, followed by the Korean War. So, many of them, as a result of that, could not go to college. But they were well taught and well prepared.

With the background in math I received from the Franklin School, I was able to be a weatherman in the US Marine Corps. It's a very complicated course. Stanford University told the United States Navy that they couldn't teach it in less than four years. And they [the military] taught it in twenty-two weeks. I was able to succeed in that course because I had a strong background in math and science. Mr. McKeeby was my high school math teacher.

And we had a strong [history department]. Problems of American Democracy with Miss Harden, who still lives here in town. She was a person who helped us become aware of

Figure 79. Since 1983, the former high school building has served as the Franklin Elementary School. (Photo: Carrie Papa)

what was going on in the community, in the state, in the area around you. We had great teachers who cared about the students and made sure that you learned. And they were very strict with the discipline. I always thought it was an excellent school.

While many children of miners did go on to an advanced education, college was not a goal for the children of most mining families. In fact, former Sterling Hill employee Wasco Hadowanetz reports that parental objectives for children in his family were just the opposite of an advanced education.

There were no expectations. When you look at our family, only two of us went through high school. What happened was, they usually quit after eighth grade and either went to work or had to work at home. There were very few [children of miners] that went on even to high school, let alone college, when my brothers were growing up. There was, I remember, even pressure on me not to go to high school. There was pressure from my father and mother both [for financial

Figure 80. Along with several other narrators, Wasco Hadowanetz spoke of the influence Franklin teachers had on his life. (Photo: Carrie Papa)

reasons]. After grammar school, we just thought we were going to quit and go to work.

I went to grammar school here in Ogdensburg. When we went to school, I remember, whoever had a car, the kids would stand on the running board and ride to school that way. We had to walk otherwise. This was in later years. But we walked to school even if we were up to our hips in snow. No school buses.

Then I went to Franklin High School and graduated in 1948. Now, I had no thought of ever going to college, but Mrs. Phillips — her maiden name was McKeeby — she talked me into going to Newark College of Engineering. I had a job in Newark. I worked during the day and went to college at night. But I just couldn't manage so I had to quit after 1949.

I went in service during the Korean War, 1951 to 1953. I was able to get on the GI Bill. I went back to college and went one year of day school. Then I had to quit because I didn't have the money to continue. So that was the extent of my college education. Actually, it was one year of night school and one half year of days. Later on, 1972, I took some night courses, and I studied for the state professional engineer's exam. I passed parts one and two, which were the equivalent at that time to a college education. So, with my experience [at Picatinny Arsenal], I put in for changing my title from engineering technician to engineer so I got my title as engineer.

~

The number of times that specific Franklin High School teachers were given credit for helping students to continue their education was impressive. Ann Trofimuk, valedictorian of her eighth grade graduation class and salutatorian of her high school class, acknowledges a favorite teacher. Ann speaks of the bias that existed against educating women and also considers the possibility that the Zinc Company may have wanted a vocational school in order to train laborers.

> Maybe in the beginning it was that, if you go back to the twenties or whenever the school [was established]. At that time, were schools set up with the idea of having children go to college? I don't think they were, basically. You stop and you think about it, not every Zinc Company management team's children went away to prep school and everything. You think of the Seips. They went through the system. And a lot of others.
>
> I think it was to their [the Zinc Company's] advantage to help encourage a good school system. They wanted the children of even — what would you call it — middle management or first level management to get good educations. It was to their advantage. I don't think anything was done to discourage young people so that they would stay and work in the mine, I don't think so. I would say not.
>
> I thought the Franklin school system was good. I learned there. It helped foster my love of learning. I just wanted to keep learning. I wanted to go to college. Now, as a miner's daughter, how was I going to go to college? One of my most wonderful mentors was Miss Titsworth, our English teacher in high school. She was a graduate of what was then NJC [New Jersey College], which is now Douglass. She was a moving force in my going to Douglass.
>
> I knew that the man who managed the Zinc Company store, Abie Wright, was in Kiwanis. So, having worked there [in the store] for several summers, when I was looking into what assistance was available, I went to him and asked him, "Mr. Wright, I understand that the Kiwanis offers scholarships. Could I apply for it?" And he told me that they didn't offer it to girls since girls probably wouldn't work when they grew up and were going to be married and such. So they offered it to boys. Thank God that I got scholarships from other sources, the state and the college itself, so I didn't need it. But, at that time, I suppose that was the attitude. I just

say "God Bless Miss Titsworth!" She was a moving force in my going to Douglass.

~

More than one of the narrators confirmed that marriage was the primary goal and expectation of most young women after high school. Laura Falcone reports that her fundamental hope after graduating from Franklin High School in 1947 was:

> . . . to get married! All I wanted to do was get married. [laughter] It never occurred to me that there was any hope of going on to college. It was an absolutely foreign idea. The concept of children going to college in our circumstances was vacant. Gone. No such thing. I don't ever remember my parents discussing college. They may have said how nice it would have been, but the idea was totally beyond us financially. It was just not thought of. High school is what most people expected for their children.
>
> Except for the owners of the mines with their children. They were automatically expected to go on to college. Because of the Zinc Company, all of their contributions, we had a lot of advantages in this small town. Our schooling was exceptional. I really believe that. We got into the Middle States Association earlier than other schools around just because it was so superior. Of course in later years on reflection you think, "Well, obviously, you had these very high paid engineers and head employees of the mine with their children in town. And they certainly wanted good schooling for them." But everybody was taught in the same school. There was no differentiation. We had a fantastic library. And the schoolteachers were beautiful, hard working. We had home economics in our school. We had sewing or cooking when we got to eighth or ninth grade.
>
> We [older and younger children] were all together.

Figure 81. Narrator Laura Falcone in 1947. "Our schooling was exceptional. I really believe that." (Photo: From the author's collection)

A Mile Deep and Black as Pitch

Today, the new educational thinking is that you can't have little kids with big kids. My goodness, they'll be corrupted. Whereas, in that day, the bigger kids set the ideal for the younger children and took care of them. And the seniors were far, far above any other class. They were a good influence. Our high school graduation was held outside under that wonderful old oak tree. It was beautiful, absolutely beautiful. And then, as the town got bigger and things got larger, it was moved indoors to an auditorium, which took away a lot of the beauty.

In fact, because I graduated at sixteen, I went back to high school for another whole year called my PG, post graduate year. That is when they called me into the office! In our school building, which was as modern as they were in those days, in each room there was a telephone. It never rang. I was in my PG year in this class, and that phone rang! I guess the teacher must have jumped out of her skin too. [laughter] I suppose I had heard the phones ring over the years, but very, very rarely. She answered and I was to go to the office. "Well," I thought, "How in the world could I have to go to the office? I have not done anything wrong. I know I couldn't have." Scared to death.

When I get to the office, our principal, Mr. Hollobaugh, was there and said that different teachers, Mrs. Cunningham and George Zigler, had recommended me for a school, a higher school. And that his friend was a vice president at Ryder College. And that if I thought that my family could get together and get up half the tuition, they would provide a half-tuition scholarship. And would see to it that I would have housing and employment to pay for it while I was there. So it would only cost my family the half-tuition to get me there.

Everybody in the family was so thrilled. They all pitched in. My kid sister worked in the movie at the time selling popcorn and candy and she would send me one or two dollars every single week to live on, which was a thrill because I could buy the things I needed. Then my bigger sister would send me money. Every once in awhile I would get five dollars. Even Nora, who was married by now and having a very difficult marriage, would send me a dollar in an envelope every once in awhile. So I went to college, graduated, and then got married!

The fifty-fifth and final commencement of a Franklin High School graduating class was held in June 1982. Today, the building is used for elementary school children; students of high school age attend a regional school.

1. US President's Commission on Coal, *The American Coal Miner*, 181.

2. Thurner, *Calumet Copper and People*, 57.

He Bowled and I Knit

The social life is centered around the Community House, which is equipped with game rooms, kindergarten, bowling alleys, library, etc., for the use and enjoyment of all citizens and societies. A public park, a public swimming pool, an athletic field, good roads, street lights and water supply, are all factors that make Franklin Borough a thriving, contented and happy community.

Franklin Borough Golden Jubilee

Probably no other aspect of life in Franklin was mentioned by the narrators as often or with as much affection as were the recreational facilities provided by the New Jersey Zinc Company. Remembering happy times and forgetting difficulties seems to be a general tendency of human nature. Mining families — whether in New Jersey, Appalachia, or the western states — remember with pleasure leisure activities and entertainment, no matter how hard the times were.

Since most company towns were remote, the companies often provided many recreational facilities. Although there was considerable variation in what individual companies furnished, a town baseball team with its own baseball field was found in even the poorest of communities. This favorite American sport did not represent a big financial outlay by the company, but it generated immense goodwill. Another investment in goodwill was a company's support of a volunteer fire department. In turn, the fire department sponsored dances, parades, community festivals, and other social activities. The Neighborhood House was the most often mentioned public service facility maintained by the Zinc Company. From the Junior Canteen dances that were organized for teenagers during the Second World War to the bowling, ping-pong, and billiards tournaments that were held during the winter months, the beloved "Nabe" remained the center of social, cultural, and recreational life in Franklin. Not only did the Neighborhood House offer a meeting place for various organizations such as scouts, the women's club, and sports leagues, it also provided housing for several school teachers and offices for the town's health and social services departments.

The number of companies that built community halls, tennis courts, swimming pools, parks, libraries, and playgrounds was far more limited. Because the Zinc Company provided all of this and more, it received high praise from the narrators for its recreational programs. Despite the various complaints about the company, all of the narrators were in agreement that the leisure time activities it provided were excellent. It was a widespread belief that living in Franklin during the Zinc Company era was a wholesome, good experience for everyone, but particularly for children.

As in other rural towns in America before World War II, much entertainment in Franklin was simple and homespun. Pastimes for children often were as unpretentious as picking wild flowers or fishing, playing house or baseball, and swimming or sledding. Paralleling the opinions expressed in the Franklin interviews, the narrators in *Appalachian Coal Mining Memories* emphasize that people with little money created their own fun and entertainment. Both groups of mining families also stress that hard work and simple pleasures contributed to worthwhile family values, which included a strong sense of community.

Speaking about the coal mining life in the New River Valley of Virginia, June DeHart observes:

> *I guess coal minin' makes you a humbler, better person in the way that you're not selfish. You share, and you're not used to a lot . . . . I think most coal-minin' families feel that way. And I think that's why they all*

like to work together — because they share their experiences.[1]

People working for the same company, sharing similar values, and participating in the town's social gatherings reinforced the close, cohesive nature of the community. Ann Trofimuk expresses similar thoughts in speaking of growing up in Franklin.

> I can't remember being taught things, but obviously, I must have absorbed things. It's so far back to think, "How was your sense of values instilled?" I can't remember my mother and father saying specifically, "Don't hurt anybody." "Don't do bad things." "Be a good person." "Try to be kind." The basics of moral living and everything I absorbed from my parents, church, school, people, the town. We were all in the same position. I think a lot of people built good lives for themselves as successful human beings because the Zinc Company provided a foundation, the roots to go on and build on.
>
> I sometimes have a tendency maybe to look at the world through rose-colored glasses. Obviously, Franklin is different now, but still there are those recollections, those memories of what it was. And you and I know that we had special times.
>
> I mean, as children we made our own entertainment. I remember you told marvelous ghost stories, right? And playing. And going to the park. I mean, the Zinc Company provided all of that, right? They provided people to supervise crafts. They had games available. You could go play croquet if you wanted to. They had the tennis courts there. The concerts on Thursday. The men from the borough would come and put out all those lovely green benches, right? The Franklin band concerts, which I'm sure they subsidized. Remember how they maintained the park grounds?
>
> Now, you and I know how marvelous that library [in the Neighborhood House] was, small that it might have been. But it was there. And the pond in the summertime. We had the swimming. We had the recreational activities. I'm sure they [the Zinc Company] paid for the lifeguards, the equipment, whatever had to be done.
>
> One of my fondest memories also is that there were beginnings to ends, and ends to days, and the weekends, and Sundays were special. I remember when the day ended; my mother would take off the work apron and wash up after everything was done. Put on a clean apron, sit down, talk. The ladies in the neighborhood would talk. Of course, in the

wintertime, the day ended, and the radio would go on. The stories would go on. My mother's favorite newscaster was Gabriel Heater.

I found that everybody seemed to get along. I mean, if there were differences, it was just the human being, human nature. Each nationality had its own society. The president of the Russian Society decided at some point it would be nice to have an orchestra. I played the mandolin. We had a couple of violinists. We had a guitarist. We met in the old poolroom given by the Zinc Company for use as a sort of mini-Neighborhood House. This is where we met. And we had this orchestra.

Looking backwards, you can still see that there are still things that remind you of the Franklin that was. My recollection of growing up is that we felt that it was a marvelous time to grow up in Franklin. There were so many special things. Franklin has changed drastically. I think we all realize that. It doesn't have that feeling that it had. Maybe there is still a little sense of people trying to maintain a sense of community. But it's impossible to bring it back to the Franklin that was.

As Ann mentioned, the town library and the band concerts, which the Zinc Company either provided or supported, were sources of enjoyment for area residents. Such company-spon-

Figure 82. The Franklin Band. Throughout the summer, weekly concerts were given in Shuster Park. (Photo: Courtesy of Franklin Historical Society)

Figure 83. The library maintained in Franklin by the Zinc Company was used extensively. (Photo: *Zinc Magazine;* Courtesy of Ogdensburg Historical Society)

sored programs brought culture to the masses and filled a very real need for thousands of people. Whatever the premises for mining company largesse, residents were happy to have the amenities provided. Bill May remembers a lot about growing up in Franklin and the simple good times that were a part of his childhood. Bill reminisces:

> I'll tell you, I've got a lot of memories. Sometimes, I'll walk from here. I'll go down across the highway, down Buckwheat Road. Down around the pond. And I'm looking, and gee, I can remember this place and that place and how this place was in the summertime. Hundreds of people. Kids. You could be uptown, and you could hear the sound coming from the pond. Oh, it was something else. The beach at Franklin Pond — when the pond was open for swimming years ago — was packed. It was like Coney Island. You could hardly find a place on the beach to sit down.
>
> I always looked forward to the Fourth of July celebration at the pond because that was a big event. We had swimming events, diving events, watermelon scramble. Oh, gee, it was something else. And we had all kinds of stuff to eat. As a kid, that's what you're looking for. Ice cream. Soda. Hot dogs. Hamburgers. Then you had the Hungarian Presbyterian

Church, and they would have the kolbasi and sauerkraut and cabbage rolls. One year, when I was a little bit older, we had a Franklin Sports Association, and we had a steer roast. We had to stay there all night, the night before, to keep the spit going with this big Black Angus steer on the spit. I can remember people saying, "My God, I can smell that up town." The fellow there would cut the meat right off the steer, put it on a roll, right? And the people were just lined up. That was the biggest draw we ever had down there. We stayed up all night. I left there like seven [in the morning], met the next bunch going down there and the guy says, "Wear your softball uniform, we gotta be in the parade." This was the Franklin Sports Club. We had teams playing in the softball league. Oh boy, I had been awake for I don't know how many hours. "Oh, okay." I got a chance to ride in the parade. There used to be a Chevrolet dealer in Franklin called Cox and his daughter had a Corvette sports car. So I rode in a sports car with her. [laughter] Then it ended up with fireworks. The place would be packed down there.

The Presbyterian Church would put on a pasty sale and they'd sell hundreds of pasties. This was separate [from the Fourth of July]. Pasties are a Cornish dish. They would have an annual pasty sale, and, my God, they sold them by the hundreds. These women worked in the kitchen of the Presbyterian Church making these pasties. I haven't had a good pasty in years. Of course, my grandmother being English, she made them all the time. I can remember in work, the fellow would bring two pasties in his lunch pail. One for breakfast. One for lunch. Those pasties were as big as your shoe. That was his main stay.

Another thing that I can remember is Father Stephen Dabrowski. He used to have a one-week fiesta down at the Franklin Pond. The Catholic Church would sponsor it. Gee, that was tremendous. One night, I heard that Father Steve was going to get a celebrity to be there, right? The story was that he went knocking on the door — who is that singer that lives in Montclair — of Perry Como. Knocked on the door, and he answers, "Father, can I help you?" "Could you find a way to come up to Franklin? I have this fiesta." "Gee, father, I'm booked. But I have a booking agent," he says, "Let me see who I can get for you." "Well, thank you very much." So who do I hear singing down there? Vic Damone! That summer his big song was, "On the Street Where You Live." I heard Vic Damone singing this, so I walked down there and saw him on the stage. Oh, boy. And he had Buddy Hackett, the comedian, besides. Father Steve, he was a great guy. Yup. He even played on the Franklin Miners football team. He was a priest and he was a regular guy. A regular guy.

Also, I remember down at the Franklin Pond in the wintertime watching them cut ice for Littell's big icehouse. Watching how they did it, putting it on the conveyor and all. Then going there in the icehouse in the summertime, when it was boiling hot outside. Then go in there, oh, it was nice. Chip off a piece of ice, you know, and that was great. Today, you give a kid a piece of ice, they'll say, "Get out of here. What are you crazy?" But, for us, that was a big deal.

Schuster, he was one of the big shots in the Zinc Company. They named Schuster Park after him, which has since collapsed and caved in. The bandstand is still there. I think it was Thursday nights, everybody in town would be at the band concert. The park was packed.

And the movie theater, I really miss that. I think a lot of people do. We would go to the movies a couple times a week, at least. A weekend or a Wednesday night double feature, that was the big thing. I'd go with my father on a Sunday evening or maybe a matinee, depending on what was playing. Well, you went regardless of what was playing, you know. Certainly, life was simpler in those days.

Christmas time, I can remember working at the [Zinc Company] store, which was nice. The boss, this Elsworth Wright, was a hell of a good man. He made like a little town on a big piece of plywood and we had it in one of the display windows at Christmas time. We had this all fixed up, and we sold Lionel trains, which were a big thing years ago. We would get all these kids after school with their noses right against the window. He'd come over to me, if I wasn't doing anything or if I was working at the counter there. "Billy, go run the trains for the kids. We've got quite a group out there now." So I would go over there and run the trains, start it and stop it, and boy, the kids were bug-eyed looking at this and that.

I don't know, but around Christmas time, everybody was happy-go-lucky. Whether they had a lot of money or they didn't, it was just something about it. We had the upstairs [at the store] and normally we sold everything up there from soup to nuts in the summertime. We had lawn furniture. We sold mattresses, springs, linoleum, window shades, venetian blinds. So a lot of this stuff was taken out, of course, in the fall when the season's over. We set up long tables. We got all this Christmas stuff in we ordered, and we'd start setting that up for Christmas. People came into the store then. They wanted to see our display upstairs. This was like Santa Land. There was nothing in town like this. In fact, this Elsworth Wright, all on his own, had this big display on the roof. It was Santa's sleigh, which was a real sleigh that he got from someplace. Then he had these reindeer, figures all cut out that he did himself and painted up with spotlights on it. This was on the top of the building so you could see it all over. This was, boy, big time now. It was great at Christmas time. Hell, I can remember Christmas Eve. We worked up till ten o'clock in the evening. People still coming in buying things. It was a very nice feeling and you knew everybody. Hell, I got to know everybody in town. A lot of friends. A lot of memories.

Because entertainment was so much simpler and less sophisticated then, holidays tended to take on heightened importance. Various holiday events, such as the parade at Halloween or the fireworks on the Fourth of July were mentioned by most of the narrators as being significant highlights of the year. Christmas probably was the best-remembered holiday. For Evelyn Sabo even a small thing like Jell-O is recalled with pleasure.

It maybe will sound kind of strange, but I remember very vividly that the only time we had Jell-O — which was red or green Jell-O — was at Christmas time, and that was a BIG treat in those days. Mom would make me Jell-O and we'd have it for Christmas.

There was at Halloween always a Halloween parade. We would get dressed up in old clothes like an old woman, or blacken our faces. Whatever, we had around. Buying a costume was unheard of. We created whatever it was that we were going to wear. We would always look forward to the Halloween parade, rain or shine, because at the end of the parade, we always got free cider and doughnuts. That was the thrill of the year, getting our cider and doughnuts.

Then after, I remember Herzenbergs would break up all of their chocolate bunnies [from Easter] and put them in bags and give them to the kids for Halloween. We always looked forward to going to Herzenberg's Drug Store and getting our free bag of chocolate Easter bunnies. I remember being thrilled to be able to go.

Evelyn's sister Laura Falcone remembers several aspects of childhood in Franklin and confirms her sister's delight about the Christmas Jell-O.

Christmas. Ah, Christmas. We always had red and green Jell-O, and that was absolutely lovely. [laughter] It was made in two regular size cake pans and cut into diamonds. Then you would have spoons full of diamonds of red and green. And we always had a turkey. I think for both Thanksgiving and Christmas. One year, Daddy took a chance on a turkey, and he won! I'll never forget that year! Mom cooked both turkeys and each of us kids could have a leg. We called it the drumstick.

Also, I remember the ribbon candy. Mom and Dad always ordered from the Sears catalog a pound, or whatever the package size was, of the ribbon candy as well as a five-pound box of chocolates. That was probably the only candy we had for the year. We wolfed that down in a week. The four kids and the parents, of course. But, that was a huge treat, that box of candy. Both Thanksgiving and Christmas were very special holidays. I think mostly because of the turkey and the candy.

Another thing about Christmas were the lights on the two little pine trees that were in front of the Zinc Company mine entrance, next to the company store. This was like the front gate of entering the Zinc Company. Every Christmas, these two small pine trees were covered with colored lights. They, and there was another little tree at the bank, were the only decorated outdoor trees in town. Walking down to see the lights and to look in the Zinc Company store windows was a highlight of the Christmas season. That and the candy.

Oh, one other time we got candy was at Halloween. This was another holiday that was very important to us. There was a parade. One year, a hay ride. Doughnuts at the end of the parade. And dressing up, of course. Mostly we went as hobos. I think we all did, but maybe I am only remembering my own costume. I used to daydream over the costumes in the Sears Roebuck catalog, especially the ballerina and the fairy princess. Of course, the idea of actually having such a costume never ever entered my head. It was so completely out of the question, just a daydream. As I said, my costume was a hobo, which meant putting on a pair of Daddy's old pants, tying a bandana on the end of a stick and carrying that over your shoulder, and putting coal ashes on your face to look dirty. For whatever reason, we thought a hobo had to have a dirty face. Anyway, that was our costume. Once we got it all together, we would go down to Herzenberg's drug store and get free candy. You had to be in costume to get the candy. What Herzenberg did was wrap up their old Christmas, Easter, Valentine's Day, any old candy they had, and hand it to the kids that came in. Getting the free candy at Herzies was the big thrill of Halloween, especially if you got some chocolate in with it.

Radio also was a big part of our entertainment. We kids used to sit on the floor clustered around the radio, which was a big cabinet affair. The two favorite programs that I remember were *The Shadow* — "Who knows what evil lurks in the hearts of men? The Shadow knows," — and *Let's Pretend*. Then there was *Mr. Keen, Tracer of Lost Persons*, and *The Lone Ranger*, "Hi Ho, Silver, Away." Later on, there was the *Sixty-two Dollar Question* with Groucho Marx, and the playhouse theaters. I can't recall the names of them, there were two or three. One might have been the Lux [soap] Playhouse.

Anyway, I did hear the famous program with Orson Wells and the invasion from Mars. I remember that Daddy was there listening to it too, and wanted to take the car — we had a 1936 Ford by this time — and drive toward the city to see what was going on, and Mom wouldn't hear of it. Daddy and I used to sit in front of the radio for hours listening to ball games. Poor Daddy, he always talked about going into New York to see a ball game. Of course, he never did.

I remember playing ball with Mom and Dad in the park. The park was just across the street from our house on High Street. It was a marvelous park with big, beautifully maintained flowerbeds, gravel walks, tons of trees and benches, and a big play area with swings. They also kept board games, balls, jacks, things you could borrow to play with. I especially remember Parcheesi. There was always someone there to organize games. One time they had a scavenger hunt that I thought was just wonderful.

The Zinc Company also maintained a pond for the townspeople. They had instruction for swimming every year. And on the Fourth of July at the pond, they had all kinds of

Figure 84. The daughters of miner Paul Moore. Part of the Fourth of July celebration at the pond was a kid's beauty contest. "Mom made us enter it each year. All four of us. . . . Well, none of us ever won." (Photo: From the author's collection)

activities, games, contests, races, and so forth. A greased pole that boys tried to climb to get a dollar that was tied to the top. Also, a greased pole hung off the diving deck, also with a dollar at the end. And a watermelon contest, also greased, with two teams trying to bring it to shore.

And a kids beauty contest. Mom made us enter it each year. All four of us. I remember Mom telling us, "Now put on your French airs." By this, Mom meant we were to put our hands on our hips and sashay around the circle looking very sophisticated. At home, Mom would demonstrate how we were to walk with the French airs. Well, none of us ever won. But, for me, and I imagine my sisters too, having to be in

Figure 85. The Franklin Theater was a popular source of entertainment. (Photo: From the author's collection)

that contest was absolute agony. But the rest of the day was marvelous.

Once a week, Thursdays I think, we would have the concerts. The benches would be set up, and the band would be up on the bandstand. This was sort of up on a hill so it could be seen from all around. It was gorgeous, a railed cement area, maybe twenty or thirty feet across, where we could go roller-skating. When the band wasn't there, that made a delightful skating rink.

We had a movie too. As a teenager, I rarely missed a movie. We went as often as the movie changed, which was probably three times a week. I think this was typical of most of the children in town. I was in high school and I remember walking to the movie about a mile or two away from our house. Totally safe. Totally alone.

The patriotic holidays too, were big events. The Fourth of July at the pond, and Memorial Day — we called it Decoration Day then — we all carried a little flag and went to the cemetery. They took us in a school bus. After that, there were speeches at the flagpole on Main Street. We had so many advantages in this small town, thanks to the Zinc Company.

All the community activities, all the things people did together, all of this feeling of being carefree and secure developed a sense of belonging, a sense of caring for one another, and pride and values. Very nice. It was a beautiful town and a beautiful childhood.

Many narrators emphasized the community's values and interests, and patriotism and pride. Former member of the Franklin Hospital nursing staff Genevieve Smith mentions Memorial Day as being a special time with a lot of community involvement. Genevieve says:

> [Franklin was] a wonderful place to grow up. It was a paternalistic town all right, with the company store and all of that. Growing up in this town was carefree. I suppose it [a sense of pride in Franklin] was just the fact of growing up here. I didn't have a care in the world. Now, we were far from being a wealthy family and my mother and father had to struggle with everything that they had. They started with very little. But I was never aware of the difference between me and anybody else who might have had more. That was never, never brought up.
>
> On Memorial Day, I remember they would get the kids all ready up at the school. They would have the buses up there and they would pile us all in the buses and then take us over to the cemetery so that we could march around. We would parade with the Legion and the band and everything. The kids all had flags that we would carry. You would wind up at the flagpole. It finished up there with a speech. I thought that was nice.

Here Steve Revay remembers how the Neighborhood House fostered a sense of community.

Figure 86. Part of the 1938 Sussex County Firemen's Association parade coming down Fowler Street. (Photo: Courtesy of Franklin Historical Society)

Figure 87. The Neighborhood House on Main Street was the social center of Franklin. (Photo: *Zinc Magazine;* Courtesy of Ogdensburg Historical Society)

The Neighborhood House was great. People used to congregate there. They'd go play cards there. Bowling was free. Shoot pool was free. The band practiced there. They had dancing classes there. Anything that was going on would have its origin in the Neighborhood House. They had the nurse lived there. Took care of the babies. Miss Pancoast, she was the head of the Neighborhood House. The Zinc Company furnished all of that. The best thing they did around here was that Neighborhood House.

And you see this house over here across the street? [Mabie Street] That was a Neighborhood House too. That was a poolroom for the Hungarians, Slovaks, you know, whoever wanted to use it. All this had in it was the pool table and chairs for playing cards for the foreign element, the Hungarians, Slovaks, Russians. They used to come in here and play checkers, play cards, shoot pool. Every night they had a man here all the time running this. The Zinc Company paid for all that.

In addition to listing some of the ways in which the Zinc Company helped the community, Robert Littell mentions how the Neighborhood House played a part in his youth.

> We had a great time growing up in Franklin. It was a good place to live. We had a Neighborhood House where people could go and have fun and recreation. We would go there after school. We walked through Shuster Park to get to the "Nabe." We'd play ping-pong, pool, cards, and bowled. Friday nights, we always had the canteen dances. A lot of kids stayed out of trouble because we had a place like that to go to. I used to go there on a regular basis, after school.
>
> The Zinc Company was more than generous. The baseball diamond at the high school was made to the exact specifications of the baseball field at Yankee Stadium. They had the engineers go down there and check all of the measurements. We had the same kind of clay and sand for building the diamond. That is incredible. They built the bleachers for the baseball and football fields, with coverings over them so the sun and elements were blocked out. I think they did a lot for Franklin.

Along with parks, the Zinc Company maintained the Marshall House, a community center for the benefit of their employees in Ogdensburg. Wasco Hadowanetz tells about the center and his memories of childhood recreation.

> The Marshall House, that was right off of Plant Street. Right by the mine. There was a Marshall who was one of the early superintendents of the mine back in the 1850s and that was named for him. There was a Neighborhood House in Franklin too, which was recently torn down. They [the community houses] were a place for scouts to meet, women's clubs to meet. I remember when I was growing up we had community parks. There was a little park right across the street from the mine. But by the Marshall House was a park where we used to play football and all kinds of games there.
>
> In fact, one of the supervisors from the mine would be serving as a moderator [at the Marshall House]. He'd let the kids do their own running of their meetings, but he'd sit there as a sponsor. Then, later on, Sid Goodwin moved into town. He was the only president of the New Jersey Zinc

Company who lived in Ogdensburg. He would involve himself with the church, the Presbyterian Church.

We also had a Community Club that met in the school. We had boxing tournaments between us and Franklin. That rivalry was not only in boxing, but baseball and football. Then, later on, of course with the rest of Sussex County, twilight meets and things like that. In those days, it was just between us and Franklin.

I remember at night walking the railroad tracks to Franklin to go to the movies. I remember every shadow you saw, you thought it was somebody after you. I wasn't really scared, but I remember walking all over at an early age.

We called the night before Halloween ticktack night. The reason for that was, as I recall, we would take empty spools from sewing thread, cut notches in these spools, and somehow tie a string around them with nails on them, and then rap them against windows and scare the daylights out of people in the homes. But some would do things like tip over the outhouses — sometimes with somebody in them. [laughter] Or take the bench down at the park and raise it up on a flagpole. Things like that.

They [the Zinc Company] built the bowling alley too. They used to have teams from the New Jersey Zinc Company facilities in Palmerton, [Pennsylvania], Gilman, Colorado, the New York office, and down in Virginia, Austinville. Then, they would have tournaments at these various places. My brother bowled on the Sterling Hill Mine team. Residents could bowl there also. I used to set pins there. I used to get a nickel a game for setting pins. They weren't automatic pinsetters. You had to put them up pin by pin.

Ogdensburg resident Clarence Case, a miner at Sterling Hill for several years, also speaks fondly of the Marshall House.

The Zinc Company provided a bowling alley run by the company. They also had a building down here called the Marshall House. I used to play there when I was a kid. Used to have meetings. Used to play games. Just a night out for young kids. There was a Mr. Sekelski that was the supervisor, and a Mr. Charlie Brown too was involved. The Zinc Company did provide different things for the people in town. That was great. Growing up in Ogdensburg was great.

Two of the narrators, John Baum and Harry Romaine, mention the Zinc Company picnics. John was part of the management team and recalls the picnics given for management.

> The company had picnics. The ones I went to were for salaried personnel. They were pretty high-class, as a matter of fact. The meat was tenderloin. There were clams on the half shell, there was copious food, and the beer flowed like water. It was a very pleasant affair. Down at the rifle range in my time. I'm sure there were [picnics] for the miners too. They were quite likely to be up in the football field or something like that.

The picnics that miner Harry Romaine remembers were held in Shuster Park. Harry explains:

> This was all paid for by the Zinc Company. Up on top of the hill where the firemen have their park now, used to be a Zinc Company park. They even paid men to take care of that park. The Zinc Company used to put on what they called the Zinc Company picnic twice a year. Everything was free. Used to go get our ice cream cones. Now, that's what I mean. That's what the company done for the people.

Many of the narrators who knew Franklin during the heyday of the Zinc Company and who continue to live in the borough, feel nostalgic when thinking back to the earlier days. As Bill May says, "It was all Zinc Company. Oh, it was good. It was good, and now everything is gone."

1. La Lone, ed., *Appalachian Coal Mining Memories*, 107.

Part Three

The Legacy and the Future

Although the Zinc mining industry in Franklin shut down in 1954, collecting mineral specimens continues at the Buckwheat Pit and Trotter Mineral Dump, which contain thousands of tons of extracted ore and tailings.

At the Sterling Hill Mining Museum in Ogdensburg, a new development is being planned. It's successful hands-on Rock and Discovery Center will be expanded to include a laboratory experience . . . for a first-class teaching facility.

The New Jersey Herald, March 7, 1998

Closing of the Franklin Mine

*The passage of the Franklin Mine is more than a
milestone in the history of the Company. It marks the
completion of one of the great mining enterprises of
the world.*

Zinc Magazine, December 1954

On September 30, 1954, the great Franklin Mine ceased
operations. On that date at the Palmer Shaft, the last skip of ore
was brought to the top. The unique ore body had been completely exhausted and a significant chapter in the history of mining in America came to an end. Although the miners in Franklin
had known for months that the ore was running out, it was a
time of sadness for everyone concerned when the mine was offi-
cially closed on that September afternoon. The immediate impact on Franklin was not as drastic as it might have been, however, since the Zinc Company still operated the mine in
Ogdensburg and the Franklin miners therefore had an opportunity to continue to work with the Zinc Company at Sterling Hill.
Also, the younger generation of Franklin residents, many of them
men who had experienced World War II, did not plan on working in the mine anyway. Still, the closing of a mine in a one-
industry town and the resulting uncertainty had a tremendously
unsettling influence on the community.

In the early years of mining development, isolation made it
a necessity for mining companies to establish towns and provide
housing, health care, and educational and recreational facilities.
After World War II, mining company towns began to decline
and disappear as unsteady markets developed, bodies of ore were

Figure 88. The Palmer Shaft was excavated in 1906 under the direction of Robert Catlin. (Photo: *Zinc Magazine;* Courtesy of Ogdensburg Historical Society)

depleted, management philosophies changed, and transportation and communication networks were modernized. When ore bodies became exhausted and mining companies moved on, the former company towns faced diverse and often uncertain futures; some continued as bedroom or retirement communities, some were absorbed by expanding neighboring cities, some were abandoned, and some even were transformed into viable model communities.

During the middle of the twentieth century, as company towns increasingly became economic burdens to the company, many mining corporations began to sell houses, other properties, and services to town residents. As a result, many company towns eventually were transformed into independent communities. By the time of the closing of the Franklin Mine in 1954, many of the miners had already purchased their homes from the Zinc Company. Although mining is no longer a local industry, Franklin, like a great many early mining towns, continues to exist, and its residents still appreciate and take pride in their mining heritage.

Bill May remembers his father's remarks on the closing of the mine and his own earlier lay off from the company store.

> I started in November 1946. I stayed there until the store closed. I think it was 1951. About five and a half years. That was the beginning of the end. Then the mine stayed on for maybe another year or two. I was offered a job when the Zinc Company store closed. They said, "Well, we'll give you a job in the mine," you know, "So you're not without a job." My father said, "No, forget about that." He said, "Too dangerous." So I said, "No, I'll look for something else."
>
> They [the miners] knew it was coming. They were told that the ore body was being depleted. Some of the old miners said, "Oh, there's ore down there." Yes, but it wasn't profitable to mine. It got down to where the yield from the tonnage that they brought up wasn't profitable. Practically everybody in town was connected with the Zinc Company. During the war years, it was a big thing for the war effort, the mining of zinc. Three shifts. It was high priority. Everything they needed, the mines got priority to purchase equipment or whatever. But then it closed. I felt so sad for my father. He said, "Bill, I've worked in there for thirty-five years. My locker was next to these men in the change house," and he said, "I saw these men for thirty-five years, and then, all of a sudden, Monday morning comes. No more. It's all gone." I knew it was tough for him.
>
> I just felt for the fellows that had worked there for years, like my father and a lot of his friends. Where were they going to go? You're fifty years old, where the hell you going to go? I'll tell you, these places where you go to, they say they don't discriminate by race, color, age, or whatever. Oh yes. Sure. Yeah! Some of them went down to the iron mines in Mine Hill. Down by Dover, there was a Richard's Mine. I think it was iron ore. A lot of them worked until they became sixty-five or whatever it was [to retire] at that time.
>
> When the Zinc Company closed, they had a company called Eastern Scrap that came in to dismantle the plant. Took everything they could use out of it, like electric motors

or whatever. Then they cut all the buildings down, you know, with cutting torches. Take all the steelwork out and what you see today, that's what's left. A couple of chimneys now and some foundations. My father worked there for a few years, for this company, Eastern Scrap.

The thing of it was — I noticed this happened too — the fellows that worked hard every day, every day, for years, when they got out of the mine, if they lived two years, they were lucky. Lost without their jobs. I'll tell you, in the warm weather, I used to see old timers sitting around down the pond fishing. Remember the old bamboo poles? Didn't know what the hell to do with themselves. I thought, "Oh, gee, what a shame."

In the following, Evelyn Sabo remembers that residents had been gravitating toward various vocations other than mining.

The mine employed probably 90 percent of all the people that lived in Franklin. There really were very few other places to work in Franklin in those days. I think that little by little because of the war, people started wanting to get away from the mine. A lot of them started working for Picatinny Arsenal, which I did also. My husband was determined he would not go in the mine. Although he worked for the Zinc Company on top for awhile.

In twelfth grade, representatives from the Zinc Company would come to the school, take all of the boys, give them a tour of the mine to entice them to come to work for them. [My husband] got started into construction [for the Zinc Company] because he was not about to go into the mine. Then later on he went into Picatinny Arsenal, after we got married.

See, now [the young men in Franklin] are wising up and they're not wanting the Zinc Company. They wanted to get out and do other things. A lot of them wanted to go to college. A lot of them joined the service. So now the Zinc Company was having a hard time to find men to go into the mine. That's why they were trying to entice all these high school graduates to come and work for them because so many of them now weren't going to work there. Before, it was just customary — everybody that came to Franklin worked in the mine. But now the trend was changing.

Ninety-two-year-old Steve Revay had been working for the Zinc Company for thirty-three years when the Franklin Mine closed. Steve too was given the opportunity to continue working for the company after the Franklin closing. Steve explains:

> After they shut down, I worked for them for a little while in the real estate division. According to the union set-up, if they had a real estate division, I could work in it because it was in the same division as I was and I had more seniority rights than a lot of the guys in the division. So I worked in real estate for awhile. And, after that, after the real estate went out, they sent me to the hospital as a janitor. I worked as a janitor in the hospital until they shut the hospital down. Then I was out of work.
>
> When the Zinc Company shut down, they left the hospital, but that was no damn good to Franklin. The Zinc Company put in the sewer lines. They put in the water lines. That was during the early years. And the water lines and the sewer lines were no damn good. They had to rebuild the whole damn business.
>
> The borough took it over after the Zinc Company left. And the electricity, the people were to blame themselves. They

Figure 89. A Zinc Company employee takes a reading from an electric distribution panel. (Photo: *Zinc Magazine;* Courtesy of Ogdensburg Historical Society)

could have kept that power house here and they would have
had their own electricity, but they voted against it. I voted for
the Zinc Company power house and their boiler house to
stay here. Let them supply electricity for Franklin. Would
have been a lot cheaper for us and we could have sold some
of it. But, the damn fools, they didn't want it. They wanted
Dover Power and Light. Dover Power and Light gave them a
cock and bull story that it would be cheaper if they brought
in power and light.

I went to Dover, after I left here. I was making seventy-one
cents an hour here, and there they started me off with a dollar
and a half. I didn't know or I would have left here long ago.
Well, I was a damn fool. I never came back here [to work]
after that.

John Naisby, who started his career with the Zinc Company in the machine shop in Franklin, retired from the company
in 1968 as the chief of construction and maintenance at the Sterling Hill plant. According to John, in addition to offering the
town the electric power plant, the Zinc Company offered to bring
a water system to the Borough of Franklin. John believes that

Figure 90. Narrator John Naisby felt that there was a lot of distrust between the company and the town. (Photo: Carrie Papa)

Figure 91. Zinc Company employees maintain the grounds of the water filter plant that was established by the Company. When the mine closed, the Company offered the Franklin Water System to the town, but the offer was refused. (Photo: *Zinc Magazine;* Courtesy of Ogdensburg Historical Society)

distrust of the company was the main reason that the borough refused the offer.

> The experience they had in Franklin was that the company offered the township the Franklin water system free to the town. Years ago, they [the New Jersey Zinc Company] allowed themselves to be taxed to help Franklin out in order to support the town. That was a long time before I came around. Now, they offered the township the water system free to the town. This was when the company was beginning to close down.
>
> Well, they [Franklin officials] thought there was something crooked about it so they wouldn't accept it. Ten years later, they paid fifty thousand dollars for [a water system]. Could have had it for free. There's always been a lot of suspicion ever since I've been around here of the company, you know. Of course, the town figures that the company is trying to take them over and the company figures the town is trying to take them over. But I guess that's typical.

Bob Shelton, former director of the town's first bank, discusses an impact the Franklin Mine closing had on homeowners.

> An illustration of what happened when the Zinc Company went: the Zinc Company was working strong, and they went right on until they ran out of ore. When that happened, they had to close this mine. When we got this house, we paid for it and our taxes were seventy dollars a year. And the Zinc Company went out. The day they went out, my taxes went up to seven hundred and some dollars, and today it's four thousand five hundred dollars.
>
> But that's happening all over New Jersey. This happened to be an exaggerated case because you had a one-industry town, the Zinc Company. One of the difficult things we're facing is the fact that, like so many smaller towns, the taxes have become unbearable to most people. We're paying about forty-five hundred dollars a year taxes on this house and I told you when we first bought it, we paid seventy. But then it [the Zinc Company] went out and everything went sky high. A lot of these people [former Zinc Company employees] had to go out and get other jobs. They didn't all just retire. It was a very hard thing on a lot of people.

Bernie Kozykowski knew Franklin as a child because his uncle was a miner in the Franklin Mine. Bernie himself worked in the Sterling Hill Mine during the 1970s. Bernie thinks it is a shame that so little interest was shown in the preservation of Zinc Company history at the time the Franklin Mine closed.

> Their heritage is directly rooted with the mining industry. I mean, Franklin is a mining community through and through, from beginning to recent end. Unfortunately, for some reason, the younger generations and those people who have replaced those who were here before them, either from within the community, or moving in from outside, don't see any value to that very, very important heritage. Those of us, who have pursued an interest in the history, the mining, the mineralogy, and all that is Franklin, in large part, were all from outside of Franklin. I don't know why a community would want to deny such an important heritage. I look at what was here, and what, fortunately, we've been able to preserve in Ogdensburg. Franklin had the same opportunity.

The Parker Dump was located pretty much where the current firehouse is. And they built a firehouse on the Parker Dump, which was a premier mineral locality. The location of the headworks of the Parker Shaft; to me, there ought to be something there indicating what's there. It's just covered over with concrete and bush. It's almost like they don't want people to know it's there.

Toward the end, after the Second World War, the mining company saw an end to mining in Franklin. They knew what the limits of the ore were, and they stopped hiring people deliberately. When they shut down the mine in Franklin, my uncle went down to the Burg for awhile. In fact, you know how long a person's been around Franklin when they start referring to Ogdensburg as the Burg. I remember that he'd gone down there and he worked there for a couple of years, or maybe not even that long, and he just couldn't handle it. People in Franklin would not work in Ogdensburg.

You didn't have very many young people working in the Franklin Mine toward the end. They were the old seasoned experienced miners who had been there since the thirties. They weren't bringing in anybody new, so you rapidly run out. If you stop to think that anybody that worked in the Franklin Mine at this point has to be at least seventy years old, a good seventy years old. Probably quite likely closer to eighty than seventy.

~

Figure 92. Narrator Bernard Kozykowski continues to be interested in his mining heritage. (Photo: Carrie Papa)

Present curator of the Franklin Mineral Museum John Baum concludes that once the mine closed, the character of Franklin changed dramatically. John regrets that:

> They didn't have the spirit here after the Zinc Company pooped out. The town just became a bedroom community. Everybody's a stranger now in town. We used to know everybody back in the old days. You don't anymore. There's no thought that Franklin has any history that's worth remembering or doing anything about.
>
> Very few people from Franklin town come into this museum. They're all outsiders. They just sleep here. The town has pretty well had it, as far as character is concerned. People are trying to revive this town. Develop downtown and all that stuff, but Main Street has gone all to pieces. What commercial town there is now is out on the highway, really.

Former Franklin miner, lawyer, and present-day bank CEO Don Kovach adds:

> I guess the way I remember it is in the families, and structure, and security. Perhaps everybody feels that way about their childhood and teenage years and high school affiliation. During the heyday of the Zinc Company, everybody walked to work. Most people didn't have a car. That's why Franklin's laid out the way it is. All the guys would walk with their dinner pails. They would change down there so they didn't have their mining clothes on. They would walk in pairs or groups of three or four.
>
> I don't think it's ever going to reappear. It looks like the economy of the world is such that there will be very little mining in our country. There'll be some, but not anything like it was here then because we supplied the world. As a matter of fact, the Franklin Mine and the New Jersey Zinc Company took care of all of our zinc needs for two World Wars. I don't know if you remember, but I remember the lights on the fences. I remember the guys walking around with the civil defense hats. I remember the air raids. I remember the tape on the headlights. And I remember the tipple bringing the ore up out of the ground going twenty-four hours a day. You could hear that going up and down, up and down. Trains coming in and out. I can remember it so well. I mean it was a very, very active town, and the reason for that was, it

supplied not just us, but all of our allies, the whole world with zinc. Now we import it from Peru or Canada or wherever.

Probably the thing that bothers me most is that Franklin had a superior leadership group in the people of the Zinc Company executives that left. It was the foresight of some of the prominent names in Zinc Company history that [developed Franklin]. Names like Catlin, who established the hospital, McKechnie, McCann, and others. Jack Baum is another example, a renowned geologist. He worked with a fellow by the name of Callahan who was published in geological magazines, an MIT graduate. All those people, gentlemen — wives included I think — contributed to an environment that was second to none in terms of making sure that the welfare of the community was advanced through educational and cultural means. They had women's organizations that put on plays, an active Kiwanis club. The Franklin Band which still exists, would give concerts. They had lectures. The community was an advanced type of environment, I think, to what it is today.

When the Zinc Company left, so did that kind of intelligentsia. The ones that had the talent, like myself, who did not have the connections here that I had, the roots that I had, didn't come back. So to see the hospital close, and that was necessary and inevitable. To see what I thought was the cultural center of town — as it is in any community, at least from a young person's view — to see the high school leave, those things bother me. Things like that. The pond, for example, became polluted so it can not be used as a focal point for summer recreation. But that's just the times I guess.

The Neighborhood House was a symbol. It's really probably not a mistake that it was taken down because the town did not have the wherewithal from a tax base to provide for the capital improvements necessary to make it functional. Secondly, the community doesn't have the resources that the Zinc Company had to create and fund, to the extent that the Zinc Company did, a recreational program that would support a community center like that. That era was over so the building could not really be improved or maintained and devoted to what it was when I was a young man. But that's an example of what has happened. But without a strong tax base and that really was what the Zinc Company provided to the community, those things were inevitable.

Figure 93. The Neighborhood House being demolished around the early 1990s. As narrator Don Kovach said, "The Neighborhood House was a symbol." (Photo: Courtesy of Franklin Historical Society)

On the dreary afternoon that the last skip of ore was hauled to the surface and the world famous mine ceased to be, Franklin was the most advanced and progressive town in Sussex County. As it was known then, Franklin was indeed the "Model Mining Town of America." For many hundreds of people, the end of the great Franklin Mine marked the end of a secure and simple lifestyle. As Bernie Kozykowski notes, "It's kind of sad."

The Tax Fight

It took about 45 minutes last night for Ogdensburg to auction off New Jersey's last zinc mine, a landmark with a 140 year history and a million dollar unpaid tax bill.

The Star Ledger, March 8, 1989

While the New Jersey Zinc Company's legendary Franklin Mine had the distinction of closing in 1954 only because the ore was depleted, Sterling Hill Mine had the ignominious distinction of being abandoned because of a dispute over taxes.

The Zinc Company had been having labor and economic difficulties as early as the 1950s. In fact, in 1958, the company had temporarily closed the mine because of low prices for zinc on the world market. Foreign imports were selling much below the production cost of local zinc. For two years the only activity at the mine was with the Lamont Geological Laboratory. Conditions on the eighteen-fifty level were so favorable for seismic research that the original seismograph station installed in 1958 was greatly expanded during these years and came to be ranked among the best in the nation. Mining resumed in 1961, but the social, economic, and technological changes characteristic of the time continued to create pressures for the company.

Disputes between the town and the company over taxes and the assessed value of the company's assets became established and expanded over the years that followed. Hundreds of thousands of dollars were spent by both sides without satisfactory resolution for either party. The New Jersey Zinc Company felt the assessed value of the Sterling Hill Mine should have been

287

Figure 94. Surface buildings of the Sterling Hill Mine. (Photo: Courtesy of the Sterling Hill Mining Museum)

about five hundred thousand dollars. The Borough of Ogdensburg and a state tax court believed that the eight-million-dollar assessment of the Sterling Hill Mine was justified. In 1981, the New Jersey Zinc Company, then a subsidiary of Gulf and Western Industries, was sold to Horsehead Industries, which was a large corporate conglomerate with a management philosophy that was entirely different from the former mining company.

The tax disputes continued for years, during which time Horsehead Industries filed several appeals. Horsehead Industries paid its taxes in full for the years 1982, 1983, and 1984, but marked each check "under protest." This meant that if the company won the appeals, the town would have to reimburse the company a great part of the money the company had already paid in taxes and the town had already used for municipal expenses. Horsehead Industries paid its 1985 taxes, but appealed the 1986 and 1987 taxes since the mine was not operating during those years. However, the appeals for those two years were dismissed by the state tax court in 1988 because the company had failed to fulfill the requirement that at least three-quarters of an annual tax bill must be paid in order to be eligible for a tax appeal hearing.

After eighteen months of deliberation, the state tax court rendered a final decision on March 4, 1988. This decision stated in effect that the Zinc Company could not justify the lower assessment value. The final result was that the Zinc Company closed the mine while still owing Ogdensburg a million dollars. The town then placed a tax lien on the property and foreclosed, and became the mine's owner in April 1988 when the New Jersey Zinc Company refused to pay the back taxes.

The last appointed superintendent of Sterling Hill, geologist Robert Metsger, had the dismal responsibility of handing the Zinc Company keys over to the mayor of Ogdensburg on April 24, 1988. Although the state of New Jersey was a leader in environmental protection and made certain that Sterling Hill was environmentally safe when it was abandoned, the state showed no interest in preserving Sterling Hill Mine as a historic site. Today, many people feel that New Jersey was negligent in not acquiring and interpreting this site as a place important to the state's economic history.

In the following excerpt from his interview, Bob Metsger explains the implications of the corporate takeover and discusses

Figure 95. Narrator Robert Metsger had the dismal responsibility of turning the Zinc Company keys over to the mayor of Ogdensburg when the mines closed. (Photo: Carrie Papa)

the 1958 closing of the mine, the tax disputes, his work with the Lamont Laboratory, and finding a place for the mine's archives.

Back in 1958 to 1961, the Ogdensburg mine had to be closed down because there was no demand for zinc and it was the most costly operation that the company had. It was much cheaper to mine it somewhere else. People didn't understand that locally. During those three years that the mine was closed, only the shift bosses were kept on. It was during that three year period that I worked totally for Columbia [University] really. I mean, I was still paid by the Zinc Company, but practically all of my work was with this seismograph station. I did some other geologic work, but mainly that's when we built that station, which had almost fifty different instruments in it.

When we re-opened [after being shutdown for three years] we'd lost all of our miners. They'd gone to other jobs. So we had to recruit new ones. We got a lot of people from West Virginia. West Virginia had a program to get people out of West Virginia because they had so many unemployed. We sent scouts to Charleston, West Virginia, to interview people. I think we hired around a thousand or more over a period of time. But they didn't want to work. They took the jobs, and at first, we had them physically examined in West Virginia, which was a costly thing. We brought them by bus to Ogdensburg, and a lot of them, most of them, didn't want to work. Some of them didn't even report to work. They just went back to West Virginia. They were on relief. It was not just those that came to Ogdensburg. They had the same problem with those people going out to Detroit to work in the automobile industry. They would go out and they'd work maybe a couple of days. Just enough to say they had worked and then they'd go back to West Virginia and get back on the relief rolls. I don't think we got more than half a dozen, out of all those people, that really stayed with the company.

When the company was owned by a conglomerate, the conglomerate's philosophy was that you had to make money. Whereas, when it was the Zinc Company, they used to stash its profits away for a rainy day. But Gulf and Western didn't think that way. Every year, continually, they took all the profits of the Zinc Company and put them in Gulf and Western. As a result the Zinc Company kind of withered. What I'm getting at is what caused the demise of the Zinc

Company was the fact that the money was taken by Gulf and Western and their exploration became minimal.

In the fifties I guess it was, there was a tax appeal and for a long time, the way the law was then, you didn't have to pay taxes until the appeal was settled. It went for several years and it was a bad, bad thing for the people in Ogdensburg and it was a bad thing for the company's reputation. But, before that appeal went to the court, the company had offered to settle for a certain assessment. I don't remember what it was now, but the borough would not allow it. They wouldn't accept that. Well, Ogdensburg [mine] finally closed. It was on its last legs anyhow. We had only about two years worth of ore left in the mine when we closed down. You may hear people say there was a lot of ore left, but the fact that the ore body dipped into the fault, you're mining out a slab of ore and therefore removing the support by mining it out. Then the fault is just graphite, very greasy sheer zone and vertical and that provides no support at all either. So we had to leave a lot of ore to hold up the place or the whole thing would come down. That is the reason ore was left. Now, we could have perhaps gone another two years, but with the tax problem with Ogdensburg . . . . It was over a period of time that it closed down. They'd [the workers] be let go and no longer needed. Most of them took it well. There wasn't any alternative. The handwriting was on the wall anyhow.

When the mine closed, we had to make sure that it was safe. For one thing, I had to make sure that there was nothing environmentally bad in the mine. For example, down in the eighteen fifty level of the mine, we had a lot of old transformers, which have PCBs [polychlorinated biphenyls] in them. Had to get those out. We had to hire a special company to come in — actually it was Westinghouse. They went down and took the oil out of the transformers. PCBs are considered terrible for the environment.

When we were about to close down, I was hoping we would be able to turn the whole thing over, lock, stock, and barrel to the state, with all of the equipment. Make it into an historic site and an educational facility. Allow people to go down into the mine. Keep it pumped out. Can you imagine what it would be if that were truly kept the way it was with all of the machinery and everything in place? The company was agreeable to doing it [giving the equipment to the state].

They held off for something like four months not doing anything about taking the equipment out, particularly the hoisting and milling equipment.

I guess I approached the wrong person. I asked [a person with the state legislature] if he could do something about it. He wasn't interested. He didn't say he wasn't interested. He just didn't do anything. Finally, the company got tired of waiting and they went ahead and took the equipment out. It probably had something to do with politics, or attitudes of the administration in Ogdensburg, and the tax problems. That really bothered me because we could have kept the seismograph station going too. I hated to see it close. It was a very interesting place to work.

All of the records which dated back to about 1900 — I've forgotten the exact date — and to some extent back as far as 1850 were all stored in the basement of the mine office at Ogdensburg. They had been in Franklin and were moved down there. I was told by the company to get rid of the records, but I could do anything I wanted with them. So I called the state archives. This was the first place I went to. It finally was suggested to me that the very best place was the Rutgers archives. There's a mining archive out in the University of Wyoming and they wanted records from the mine. I didn't think they should go out of the state, so I kept them here. I ended up turning them over to the Special Collections Archives at Rutgers. They have very good facilities down there. It's all climate controlled.

Everything we had is there with the exception of some maps. They didn't have facilities to keep them. Those were filed with the State Geological Survey. With the recent curtailment, they had to get rid of them so they loaned them to the Haucks down at Ogdensburg, at the mine. So they are there now. But all of the other records, everything else that existed are down at Rutgers. They're cataloged. The curator and two assistants came up. I think there was two and a half truckloads of stuff that they took. They cataloged it all and it's there for everybody to see. [They could have gotten lost in volunteer organizations.] That's the reason I gave them to Rutgers because you never know. For example, the Mineral Museum, it's doing well now, but you can't tell what it will be if different people take charge of it. It's too subject to individual preferences and that sort of thing. Some people just don't care about things like that.

∽

Bernie Kozykowski cared immensely about the preservation of both the records of the New Jersey Zinc Company and the mine itself. His uncle's experience in the Franklin Mine, his youthful collection of minerals, and his own experience as a miner all contributed to a strong interest in preserving the mining history of the area. As well as discussing labor problems and the tax dispute, Bernie comments on how local history is forgotten as a community changes.

> As a community today, Ogdensburg is not a mining community. Much like Franklin, once the mine shut down, it rapidly lost its identity. When an industry leaves, you tend to accumulate housing that's inexpensive. People from other areas will move into that community and assimilate the available facilities. It happened here in Franklin. And it happened in Ogdensburg. The mining families either moved on out into other industries, and left the area, or retired. The community started to develop a new identity. People who come into the community are somewhat overwhelmed by that history and they don't understand. I worked in the mines for a couple of years so I have a sense of kinship. I worked as a miner. Once you've done that, you develop a different tie, if you will, to everything — your ties to the community, your ties to the mining industry.
>
> Things had changed in the 1950s, when Franklin had shut down, and Ogdensburg had shut down for awhile, the mining company lost most of its former experienced miners. [Then when they opened up again] they went through recruiting programs and sent all over the country looking for miners. The company was having a hard time getting help. We had a lot of what we called hard timers down there. People who'd served hard time in prison. They were the exception rather than the rule, but that happened. They always had ads in the paper. People usually went to work for the mining company when times got tough for them. They lost their jobs. They couldn't find work. They went there until times got better. The mining company supported a lot of people. Some people would be there yet today if the mining company were still in operation. That mining company could have been there for another five to fifteen years, depending upon what was happening with the reserves that they were encountering.

Who's right or wrong in this tax fight? I think it's the combination of the state of New Jersey and the Borough of Ogdensburg. I think this country as a whole is naive in thinking as little of industry as it does. I have no sympathy whatsoever toward the Borough of Ogdensburg because of their blind stupidity and their power mongering, if you will, for political purposes. You've got to understand, the real wealth of any country is in its natural resources, the extracted resources, be it mining, forestry, agriculture, whatever.

The mining company, first of all, was expending millions of dollars annually in Ogdensburg, not only in payroll, but in keeping the operation going. They, the borough, were so badly informed and they were so greedy; they thought because they had great political strength, that they could strangle this mining company. The Borough of Ogdensburg insisted that the property assessment should be over eight million dollars. And the state of New Jersey didn't step in and help in the right way. The company said, "We can't afford to operate on that assessment. There isn't enough margin to keep this operation going." The borough wants over eight million. The company says we're only worth a million. You always wind up somewhere in between.

They had gone through that process and they had arrived at an understanding. But then [one of the Ogdensburg group] barged in on these negotiations and cursing and swearing said, "Who do you think you blankety blank are? That's eight million and that's the way it is." With that, the people from the mining company simply stood up, closed their brief cases, walked out, and shut the mine down. And they, Ogdensburg hurt Ogdensburg.

Everything that the New Jersey Zinc Company had in Ogdensburg in terms of its offices, all of the equipment in the shop, all of the operating equipment, they were willing to leave all of this in Ogdensburg. They had one of the finest mining archives, company archives, in the country. Some of the most important geological records, scientific records in the history of the country were located in Ogdensburg. And the company was willing to leave all of this intact, including one hoist so they could access the mine for preservation and research purposes.

I was in the middle of this. The same damn fools in Ogdensburg, and some people here in Franklin, by the way, started shooting their mouth off, and the attorneys for the company said, "If that's the way the Borough of Ogdensburg feels about it, pull the last hoist, and close it up. Get rid of everything." They walked away from revenue that was useful to them. I don't know what their net income was from Sterling, but it was well into seven figures, probably into eight figures, on an annual basis and they lost that. The company just gave up. They said, "What's the point? They really don't want us."

As Bernie Kozykowski and Robert Metsger explain, the Zinc Company was having trouble getting miners after the mine had been closed for three years. During the time the mine was closed, Tom Sliker went to the Mount Hope Mine in Dover. It was seven years before Tom returned to work for the Zinc Company again. Here he makes observations about the new employees and gives his opinion on the tax controversy.

When I left, everything was very strict. I liked it myself. When I left, the mine had nice, good rules to live by. I came back seven years later and the mine was changed so much, it wasn't the same thing. The people didn't know what they were doing. They had people down there with long hair. You had people didn't know bees from baloney about mining and rules weren't like the rules that we had. Gulf and Western was in it at that time. People would come and go whenever they wanted. Just like today, you can't tell people to go to work. They go to work when they get ready.

The last three, four years that I was there, maybe six years, they started bringing in men from Virginia, down South. They just went down there and picked up a whole bunch of men and they were bringing them in here by busloads. Some would last two days. Some would last a week. Some would last a month. And these guys were bums. They were rummies. Drink and, oh, it was terrible.

During the last four or five years, before I went on pension in 1986, we had people talking about drugs. I know we had one man selling it, but nobody ever caught him. I don't know who he was selling it to. I didn't know of any miner that was using that stuff, but I suppose there were. Discipline was gone. The whole thing was so much different. And, of course, they could buy ore cheaper from France than we could mine it here. But it wasn't just the ore.

Ogdensburg wanted to tax the company for ore that they didn't mine yet. In that tax fight, I think the town was wrong. They taxed the company for ore that was down there. Now, how can you tax anybody for ore — you can't get money out of ore just laying there. You've got to get it out first. I think that was wrong. For myself, I think it's the town's fault.

John Kolic also worked at the Sterling Hill Mine during its final years. Like Tom Sliker, John speaks about the turnover of employees and the working conditions during that time. He too feels the borough did not act appropriately in the tax dispute. Earnestly, John declares:

Produce a valuable product for society. Getting the minerals. To me, that's fine. There are parts of the job that are physically challenging and mentally challenging. If you can cope with it successfully, you can take pride in that. Too many people thought it was a low status job. Not everybody got into it. In 1972, when I started, they had a high turnover rate. They were hiring big groups of people frequently. Like every two weeks, they would hire ten, fifteen people. Some of those would walk into the adit, see the cage, turn around, and leave. Never go underground. Others would go down, come back up at the end of the shift, and never come back the next day. Some people, it just wasn't for them. There were top-flight guys and there were very lazy guys and there was a wide range, there were all gradations in between. The best workers made maybe ten dollars an hour bonus. There were other guys that didn't care. They generally didn't like their work and just showed up to get the check.

I've heard there were drugs in that mine, but I don't know anything about it or the names of people that might have been using it. In 1972, there were a lot of long hairs there, and if they were using drugs on the outside, they probably were using them in the mine too.

There's a law, for safety reasons, that you can't work alone underground. By and large, they [the company's safety policies] were pretty good. But there were little things that would slip sometimes, such as the working alone. Now, one of the ways to interpret having a helper is if there's another guy working on the level within earshot of you, okay, who's supposed to do his work and keep an eye on you too. So he could be called your helper. Therefore, you're not working alone. Sometimes that was stretched a little too far and a person may have been hurt or an injury may have been aggravated because no one was there to help. That happened in a few cases. Especially after the layoff they had in 1980.

Prior to 1980, they had helpers and miners that worked together. In 1980, they laid off about half the work force. The way they did that was by eliminating the classification helper, and laying all those guys off. They only kept miners, drill runners, and miners. What they did now — you weren't required to work alone but they would have two miners working a place instead of a miner and a helper. By having two miners in a place, they expected each miner to be running a drill. You had to do the job yourself that the helper used to do which is hand you the steels, and in general help you move the machine around and things like that. Now you had to do all of that yourself. So you had two drills going at the same time. There, the guys were in the same place, so they could keep an eye on each other in case there was an accident. The work wasn't really that bad that you couldn't actually do it, but by the end of eight hours, you were more tired than you had been before.

My personal opinion [on the tax assessment] is that the town was hitting them pretty hard on taxes and trying to get more. As a miner who lost his job because of the whole thing, I have a bias there. My personal opinion is that the town tried to get too greedy and hit the Zinc Company for higher taxes at a time when the market conditions were bad so the company said, "We don't need this. Goodbye."

That's exactly how it was too. We come out at the end of the day and the boss said, "Don't go down to the change house, go into the office. The superintendent wants to see you." We got a lay-off notice and that was it. We never went back. No warning whatsoever. I was shocked. Everybody was.

Steve Sanford, assistant manager of the Franklin Mineral Museum, echoes John's feelings regarding the tax dispute.

My overall impression is the town got greedy. They depended, since their incorporation as far as I can tell, on Zinc Company taxes. They upped the ante one more time,

increased the tax rate on the remaining reserves and the Zinc
Company said, "That's absurd." The town took it to court.
The town won and the Zinc Company says, "Goodbye folks."

They had at least five, maybe more, years of easy mining,
easy ore production. At least five hundred thousand easy
tons left. Basically, the town, if they had kept the tax levels at
what they had been, the Zinc Company would still be
producing. They were pretty darn decent, basically, a pretty
decent company. In my way of thinking, they were a fine
place to work.

In considering the tax dispute, the narrators sided with the
Zinc Company far more often than they did with the Borough
of Ogdensburg. Geologist Bob Svecz, who worked at the Sterling Hill Mine until it closed and who today serves as a tour
guide for the Sterling Hill Mining Museum, believes that several
factors contributed to the closing. Bob feels that Gulf and
Western's corporate policy was part of the problem and that the
borough's attitude was another fundamental part. When asked
who was right and who was wrong in the tax controversy, Bob
explains the mining philosophy.

> Somebody asked me that question when I was giving a
> tour here. My opinion, and this is my own personal opinion,
> the town lost. The price of zinc at the time was the lowest it
> had been in twenty years. Now, a mining company will keep
> a mine going, operating a mine at a loss, because if they just
> close a place down like they did here, to go back into the
> mine now and repair the infrastructure would be cost
> prohibitive. It would be almost impossible to make any money
> off of it. An average mine you have fifty years of mining and
> maybe five, ten years out of that you lose money. But the
> other forty years you make money. Ideally [a mine would be
> willing to lose money for a few years and keep the mine
> going]; that's a mining company.
>
> But this company was purchased by Gulf and Western, a
> conglomerate holding company. They like to be able to show
> the stockholders a profit every year. So it comes down to that
> point and you lose the principle, the old principle of mining,
> the idea behind it. If they can't show a profit, they don't want
> to have anything to do with it. Before, the presidents of the

Zinc Company were mine engineers where they had the mining background. Now, the company had corporate officers in Gulf and Western, or when they sold it to the Zinc Corporation of America, Horsehead Zinc, or whichever, they had some kind of structure play with it.

I've heard that this mine was kept open as a leverage buyout. Then, when this place lost and they weren't making a profit on it, they let it go. But at the same time this happened, it happened that the price of zinc was so low on the market. It costs us thirty-one cents to mine a pound of zinc and we could only get twenty-nine cents on the market. We were losing money. But it might have been beneficial to operate at that loss as, I understand, a year later after the mine closed, the price of zinc went up to well over seventy cents a pound, the highest it had ever been. But the mine had already been closed.

When I started in 1971, they had an ad in the paper every day. If you go back through the classifieds in the *Herald*, let's say 1971 or 1972, you'll see every day in the paper, there was a company logo and an ad for miners. It attracted people from further off. Most of the people were not local people. When I first started in 1971, they used to take two cages down and we could fit forty men on a cage at that time. When we closed the limit on the cage — safety inspectors came in and changed it — went down to thirty and we only ran one cage to get all of the men underground.

In 1980, we were working two shifts, and they laid off half the shift. We went to one shift working underground. The price of zinc had something to do with it. The price of zinc could have stayed the same, but the cost of mining it [went up]. Everything had gone up except the price of zinc.

It was also at that time between 1980 and sometime in 1984 or 1985, everybody on the staff here had to take a 6 percent cut in pay. We didn't get that back until a year or so before the mine closed. So, let's say in 1985, I was making just as much as I had been making in 1980. But everybody on the staff stayed on. When things looked like they were turning around, we got our salary back to what it had been five years prior. We were looking at ways to cut costs. This is in 1986. They figured maybe try going to a four-day workweek, working ten hours a day.

That would have eliminated one day of start up time and shut down time. When you punch in and you're on the

Figure 96. Aerial view of the head frame and shop buildings of Sterling Hill Mine. (Photo: Courtesy of the Sterling Hill Mining Museum)

payroll, you're not actually working if you're working underground. You have to meet with your shift boss. He's going to tell you what he expects you to do that day. You're going to have to get on the cage. Ride the cage down to the level. If you're on one of the bottom levels, of course they're going to let the other guys off before you. That's another fifteen minutes before you get to your work place. Get into the work place, get your equipment set up, you're talking maybe at least a half hour. Closing would be the exact same thing. Even in the mill, the mill had to be started up in a sequence. You just couldn't push a button and have everything run or start up at the same time. You had to go in a series. And at the end of the shift, it had to be shut down in a series. So if you're talking about, I believe, seventy people at the time, you're losing time. But the problem was in the union contract. It stated if you worked more than eight hours a day, you had to be paid time and a half overtime.

So the men came up on March 27th — that was Thursday, 1986. Just before Easter break. They met in what was the mine office to vote on changing the contract to work a four-day workweek, ten hours a day. Then — I didn't even know about it at the time — the general manager walked in and more or less told them, "You don't have to bother voting on

Figure 97. The Sterling Hill Mine head frame and hoist house. (Photo: Courtesy of the Sterling Hill Mining Museum)

this. We're closing the mine and we don't know if we're going to re-open. Or when we're going to re-open. When we do, we'll call you." Of course they didn't re-open. A skeleton crew was still kept on. Everybody on the staff was still kept on. We were doing the upkeep or maintenance work, anywhere from mowing the lawns to charging the motors underground, checking the pumps. Twelve or thirteen of us.

I knew it was coming. Within the next four or five years, there was a good possibility. But, when it did shut down, it was kinda a shock. I'm looking at it from my own point of view, my own logic. The town is fighting this thing. They wanted to increase the taxes on the property and the company said, "No, it isn't worth it." The town got the idea you can tax it more when it's worth less than it was a year ago. I didn't think the town would look at it the way they did. If the company decided to close, they lost the tax revenue. And all of a sudden like that, they lost it. I think the town in the long run lost.

Up until the time when I was dismissed, everything was still in readiness to start the mine up. I was the only acting surveyor here as well as the only mine geologist. If they started the mine up, you'd need somebody that knew where the map sections were, where all the monuments and line plugs were, so they would have kept me on. But, when I was let go, the decision was made to abandon the mine. A lot of things went with the company closing before it should have.

Geologist John Baum, in very emphatic terms, sums up his feelings about the Zinc Company leaving without paying its taxes.

Yes, they did. I don't blame them! You've got a bunch of people living in a bedroom community who know nothing about the mining business, and who feel that the ore underground should be taxed whether it's going to be taken out or not. Sock it to them. They refused to understand and be sympathetic and see what it is you have to do with a mine. Ore in the ground isn't worth anything, unless you take it out. If you've got a dollar bill, that's great. It's worth a dollar. But, if you bury it and guarantee you'll never dig it out, it isn't worth anything.

Well, this ore was supposed to be worth something. But it was in the ground, and until you dig it out, it isn't worth anything. Ore is mineral that can be recovered with profit or a reasonable hope of profit. And they had no profit and no reasonable hope of profit with that stuff in the ground and the company being taxed.

The Zinc Company was gobbled up. It was gobbled up by Gulf and Western. This was the time of takeovers of companies by other companies. So today, there is no New Jersey Zinc Company. Remnants of it were sold to some of the executives who reorganized the remnants as the Zinc Corporation of America. But the Zinc Company's dead as far as I'm concerned. As far as anybody's concerned.

Thus, the New Jersey Zinc Company, which had maintained leadership in its field for over a century, came to an end. The

Figure 98. Narrator John Kolic reported, "She's going down. They're pulling the hoist. It's over." (Photo: Carrie Papa)

depths of the great mines of Franklin and Sterling Hill, with their unique ore deposits, will never be seen again. Bernie Kozykowski reports that miner John Kolic called him at the end, and in a very choked up, very sad voice told him, "She's going down. They're pulling the hoist. It's over." And so it was.

"The Fluorescent Mineral Capital of the World"

> *The hopes of many people interested in Franklin and its minerals will be realized this fall. Late last year the Franklin Kiwanis Club purchased the lot and building housing the Replica Mine Exhibit on Evans Street. The Borough of Franklin sold the parcel for a nominal figure with the understanding that a mineral museum be constructed on the site. In April 1964, a non-profit corporation, the Franklin Mineral Museum, Inc. was organized as a community project under the auspices of the Franklin Kiwanis. Actual construction should begin later in August. Every effort will be made to open the museum on October 10th, the weekend of the annual mineral show.*
>
> The Picking Table, August 1964

The Franklin Mineral Museum was opened on October 9th, 1965, by the Franklin Kiwanis Club, which had held mineral shows for several years. The mission of the museum was to preserve and interpret the mineral heritage of the area. The nucleus of the museum's collection was a small number of mineral specimens that had been given to the Franklin Vocational School by the Zinc Company, but the museum now displays thousands of mineral specimens, along with fossils, Indian artifacts, and other items, in a number of exhibit halls. One of the most spectacular exhibits is the fluorescent room where the various ores and minerals are displayed under an ultra-violet light that shows off the

Figure 99. The Franklin Mineral Museum and Mine Replica. (Photo: Ernest Duck. Courtesy of the Franklin Mineral Museum)

brilliant, extraordinary colors. The museum also provides guided tours and allows visitors the opportunity to collect minerals from the adjacent Buckwheat Dump, the mine waste pile dating back to the 1870s.

Former Franklin miner Nick Zipco was one of the original exhibitors in the first mineral show held at the Neighborhood House in 1956. Through the years Nick accumulated over two hundred tons of area minerals, and now owns the largest private collection of ore extracted from the Franklin Mine. Nick expresses his interest in minerals and his desire to have a mineral museum in Franklin.

> I lived on High Street. I lived right by where the [Trotter Dump] was, and I would come out to go to school and I would see somebody digging over there. People come in here from Paterson picking these rocks all the time. Today, they've got a firehouse there and they won't let them dig in there. But I used to see this one person in there. Nearly every week he was in there. That interested me too. What was he doing, getting all them rocks?
>
> Then, when I got in the mine, I had opportunities to go to any level in the mine. What I did on my lunch [hour], I ate as quick as I could and I went looking for minerals. Then I got involved down there in Buckwheat, taking care of the dump so that made me more strongly interested.
>
> I wanted a museum in this town in the worst way. But what the town wanted to do, they wanted to sell the dump to a guy right across the street who had a machine shop. Twice he wanted to buy it. Twice. And I stopped him. I complained too much. I didn't want them to sell it. And the third time

what they wanted to do was put the town garages in there. I fought with the town fathers for a museum for many years. Then the Kiwanis got it. They [Franklin] just loaned it to them. That's Franklin's dump. I like it the way it is now. Jack Baum's in there now. He's a Franklin man. I don't want to see a stranger to go in there and run that museum.

Like Nick Zipco, Ewald Gerstmann amassed a large and distiguished collection of Franklin minerals. Because Ewald did not approve of the Franklin Mineral Museum, he built his own museum in which to house his private collection. Here Ewald explains his objection.

I bought the property here just to put up a museum. What prompted that, at that time we had Kiwanis here. They wanted my collection as a gift. They were planning on putting a museum up. I was invited up to one of their dinners [to discuss donating my collection to their planned museum].

The building that they were going to put up was so small that it was like an outhouse. I didn't like it. That's what prompted me to buy this place. I thought, "I'll put up my own museum." I was here a year before they ever opened up. I was already established.

I had people come from all over the country. I started buying major collections. I started buying to fill this up [the Gerstmann Mineral Museum]. I bought collections from people that had Franklin minerals. I traced them down. I went all over the country. I traveled as far as California.

I mortgaged the building three times. When you get involved in something, it becomes an obsession. At the time, I hadn't got up to a complete collection. That's the motivation that people get. Every species. The best deal was when I bought the Lang Collection. He was a native in Franklin. Just the one piece ended up worth more than ten thousand. Today, that same stone is probably worth twenty thousand dollars. Some of these stones are worth more than a diamond. Anybody can buy a diamond. You can buy a diamond. What makes them so expensive? One of a kind. At that time the Lang Collection was a famous old collection, put together and written up in 1895. And the crystals. Right now, I couldn't buy any minerals because they're so high.

I didn't sell it to them [the Franklin Mineral Museum]. I

had [sold it to] a corporation, SPEX. They're a chemical corporation in southern New Jersey. I let it go for approximately [the cost of] what I had in it. Then, when SPEX paid me for the collection, I had the opportunity to have the collection go where I wanted it to go. In other words, you bought my collection, my stipulation that prompted this [sale] was that I still had the say where I wanted to put it. They [SPEX] donated it, under both of our names, to the Franklin Mineral Museum, which to this day I regret. But my collection is down there now [at the Franklin Mineral Museum], SPEX-Gerstmann.

They didn't do justice to it. Some of my friends said it should stay in Franklin. Four or five of my friends, Kozykowski, Dr. Miller. [They influenced me] to leave it in Franklin. But they didn't do justice to it at that time. Then, they [the Franklin Mineral Museum] put three additions on. Now it's straightened up a little better.

I used to be a trustee [of the Franklin Mineral Museum]. Now, I don't even go in there. I just don't like the way they handled things. I didn't make any money on it. I lost money. If I would have had that [today], I could have lived the rest of my life like a king, on what I could get. No, I wouldn't have sold it. But, if I would have died with that collection, the wife and the kids, they wouldn't know the value of it and other collectors . . . .

To me, the motivation to do this was that that's the only mark I ever made. And I was instrumental with the famous Phillips to name Franklin the *Fluorescent Mineral Capital of the World*. That was Amos Phillips that done that. Phillips Steel. *Fluorescent Mineral Capital of the World.* The original paper's up there [on the wall].

In my day, I'd set up a kid. Just last week, I had a kid in here. You give them things. They're going to be the future collectors. Right now, you're driving the future collectors off because this stuff is too high. In time, you're not going to have future collectors. Museums, there'll always be. But if they [collections of local minerals] keep going out of the area . . . . It's important for particular collections to stay in the area that they came from for future generations to see.

As Ewald Gertsmann states, former miner and mineral collector Bernie Kozykowski was influential in Ewald's decision to sell his collection to SPEX and have it stay in Franklin. Here Bernie adds to the story of how the SPEX-Gerstmann Collection came to be acquired by the Franklin Mineral Museum.

Ewald Gerstmann's very important. The Gerstmann Collection was probably the most comprehensive collection of minerals in Franklin ever assembled by any one individual. Perhaps even more so than Nick Zipco's, though Nick wouldn't want to admit it. Only because of the way Ewald did it. His collection was organized as a museum, systematically, and it was followed through. He had this little museum over there on Walsh Road. The collection's gone.

He sold the collection, my gosh, fifteen years ago — and kept it. Here's a guy, he sold the collection for one hundred and twenty-five thousand dollars, at least fifteen years ago. Then [he] was paid a stipend every year to house it in his museum, to maintain the collection for five years. So, in addition to his money [received for the collection], he received a certain amount of money from the SPEX Industries, the Mittledorfs, [to take care of it]. At the end, there was a stipulation — there was one final payment that he would have received — that he also would be instrumental in deciding where the collection would go.

Now, Ewald was always at war with the community. The Kiwanis Club here in Franklin was a powerhouse, and it was also a very divisive organization in a subtle way. Anyway, to make a long story short, the Kiwanis Club was responsible for putting together the Franklin Mineral Museum in its early years, and they deserve appropriate credit. I spent ten years in the Kiwanis Club here. You don't beat them. You join them and you work with them. Anyway, the Mittledorf family — Harriet and Arthur Mittledorf — owned SPEX Industries in Edison. They had bought the Gerstmann Collection, and the last stipulation was that he [Gerstmann] would participate in the final disposition of the collection. Now, as it turns out, the Mittledorfs weren't particularly fond of people here in Franklin. They were intellectuals and they felt people here were a bunch of fools, particularly the collector community. They were negotiating with the state of New Jersey. The state of New Jersey had said, "We will take the Gertsmann Collection." And, again, they were talking about the finest collection of Franklin minerals in existence, privately assembled. The state geologists, the state museum in particular said, "We will take your collection."

This thing was very hush hush. Somehow it leaked to me, and I said [to Ewald], "They'll stick it in a cellar, and nobody will see it." "That's stupid. You can't do that, Ewald." I said, "Is that all you want? To have your collection go to Trenton where it's going to sit in a cellar where nobody will see it?" "Well, what am I supposed to do?" I said, "Tell them you want it to stay in Franklin." I said, "Donate it to the museum on Evans Street." "Well, I don't know about that." So I made a couple of phone calls, and I said, "Look, if I can arrange for them [Franklin Mineral Museum] to take it . . . ." He said, "Well, it's going to cost them." I said, "Why?" He said, "Because if it doesn't go to Trenton, the Middledorfs won't give me my last eighteen thousand dollars." So I said, "Let me see what I can do with the museum board."

I was on the board of trustees, vice president at the time I think. Anyway, I said, "Let me make some phone calls." I called Jack Baum. I talked to Manny Honig, the attorney for the museum. I said, "Manny, we've got something special here. This thing should stay here." Manny was not new to collecting. He said, "Bernie, I agree with you. What do you want me to do?" I said, "Let me go talk to him — Gerstmann." I said, "I think we may need a special meeting of the board or something. He [Manny] said, "Look, do what you're going to do. You've got my support."

So I went back to Gerstmann and said, "Ewald, here's the deal. The Franklin Museum will take your collection. It will pay you the eighteen thousand dollars. We'll find a way." "Well, how do you know that? You haven't talked to the board." I said, "Trust me. Have I ever failed you before?" He said, "No." I said, "Fine." There are a lot of good people on the board and we'll make this work." I said, "I'll tell you what, we're going to have a meeting tonight in Manny Honig's office and we'll resolve this." We sat down and we resolved it. We saved the Gerstmann Collection for Franklin. It's in the Franklin Mineral Museum.

Former Ogdensburg miner and current assistant manager of the Franklin Mineral Museum Steve Sanford, speaks of what is involved in his work at the museum.

I take reservations. I greet folks as they come in. I help sell materials. I arrange for assistance. Keep things running in an orderly flow. Try to increase the prominence of the museum. Quite a variety of things.

The museum is not supposed to [make a profit] because it's a nonprofit organization, but they make a tidy sum, which they almost invariably sink back into the enterprise. Directors are volunteers, but they have to pay their staff. There is no money left over. But, if there is, it's usually put back into the museum in the way of exhibit cases and things like that.

There is no person more familiar with the Franklin Mineral Museum than John Baum. John has been involved with the museum from the very beginning. As curator for over thirty-five years, John speaks with authority about his work with and for the museum.

I was involved essentially from the beginning. The Kiwanis built this, you know, to put on a mineral show. And they figured because people would come to see a mineral show, maybe they would come to see a little museum. So they got the mine replica from the town and set up an exhibit there [of minerals] that the company had donated to the

Figure 100. Narrator John L. Baum has been curator at the Franklin Mineral Museum for many years. (Photo: Carrie Papa)

school, and the school donated to the library, and the library donated to the Kiwanis. That was the original museum.

I had nothing to do with that, but when they decided to start an honest-to-goodness museum, they called a meeting of anyone who was interested. I went to that and that's where they planned the present museum. I was in on the planning of that because none of them knew what a museum looked like, what the cases should be like, or where the rocks were going to come from to put in the museum. That was my responsibility.

About the Gerstmann Collection, I know Bernie [Kozykowski] was in on that. He was a great friend of Ewald's and there was some thought that that collection was going to go elsewhere. It was sold. It was purchased by a wealthy industrialist and he was going to give it to perhaps Rutgers, or Trenton, or something of the sort. Bernie persuaded Ewald Gerstmann that it should stay in town here, so ultimately we got it.

It is a great advantage [to see specimens in a museum at the original site] to people who happen to be at that site. But, for people who are living in cities, it would be better to have them in a museum in the city where they could see these things. Here, there's a great deal of benefit because we have the remnants of a mine across the street and we have the dump in which kids can dig. Most museums don't have that advantage. Children would learn the most in the museum if they paid attention because the guides are equipped to tell them about the mining methods and why they were mining here and how much money came out of this little town and so on. But children would rather collect on the dump. The adults who don't know anything about it enjoy much more seeing the museum, especially the fluorescent collection. This is the one thing I'm sure they remember. But the children enjoy the outing with all of their friends pounding on the stones.

This museum requires, but doesn't always get, a great deal of spring cleaning. Dust gets in those cases. And it's not only the glass that gets dusty, but also the specimens themselves. They have to be dusted. Some specimens have to be washed. If you have ten thousand specimens to take care of . . . . Some you can blow [the dust] off, but others are very tender indeed. It's hard for people to believe that a rock that's blasted out of a mine could be so delicate that if you just touch it, you ruin it. That's one problem, house cleaning.

The biggest problem in the museum that I can see is getting the money to come in. We're right on the edge. We never make a lot of money. But, like everything, we get more and more modern, and you have improvements. Now we've got electronic stuff that I have no use for. We're on the [Internet] and we can receive messages by the telephone and they're all typed up and whatnot. It's just all over and above my head. And what these things mean is expense! Expense! If you can sit down and send messages back and forth to California and copy things out of magazines and whatnot, these things all cost us money. The paper that this machine uses up is extremely expensive, and they buy boxes and boxes and boxes of it.

We're affiliated of course with the Franklin/Ogdensburg Mineralogical Society and the headquarters of the society is here, so that [its] announcements and stuff like that come out of all this machinery. That's one of the things. On the other hand, we do have the mineral show. And, as long as we have the mineral show, we'll be able to take in seven or eight thousand dollars a year, we hope. And that's our salvation. It's from admissions to the show up there and rentals of tables. See, the dealers have to pay us. So we get that.

The treasurer will tell you we're doing fine. The treasurer will come in, "Jack, you're being gloomy again. We came out seven thousand dollars ahead last month." I said, "Well, that's wonderful. I'm glad to hear it, but what did you do in January? What did you do in February? [The museum is closed during the winter months]. Look at the bills." The insurance is over seven thousand dollars on this place. Just the liability's seven thousand. We've got other insurance as well.

Senator Littell is there when we need him. We recently used him to help us put some weight on some other politicians. He's been very cooperative. The brown signs, the historical signs that you see on the highway are all thanks to him. For us, and for Sterling as well, he's been aggressive.

The local municipal signs that said *Fluorescent Mineral Capital of the World* are gone. We had a change in politics here in Franklin. Mayor Hodas, who was so friendly to us, was replaced by a different mayor and council, who regarded this museum as a private, for-profit operation. They saw no reason for them to go out of their way to publicize it. We offered to pay for an addition to the bottom of the present [historical] sign and there's been nothing done about it. A shame! Narrow minded for goodness sake! We're bringing business into the town. We tell them and tell them and tell them. I don't understand it.

We have close to fifteen thousand [visitors] a year. Something like that. We have a great future if we can just get expenses under control. They're working on it. I think I've awakened the people now. You see this museum isn't that old. The hopes are that by increasing our lobby in the building, we'll make our little store more attractive and sell more merchandise and have better displays in the lobby to make it more attractive, and that we'll get more schools and so on. We have Montclair State College people interested in doing archeology work here, industrial archeology. They're having a little trouble getting all the permits they need, that sort of thing, but in another year or so, we'll be having college classes coming up here. That will help the museum some because we'll be cooperating with them. Hopefully, they'll get grants and some of them will spill over in our direction. We have a great future.

Despite John's concern about the museum's future, there is no doubt that, as the years go by, the Franklin Mineral Museum will continue to fulfill its stated purpose: *to encourage a scientific interest in the young and to reveal to their elders the national treasure that was the Franklin Mine.*

A World Class Museum

Only the privileged few were ever able to step into the most unique site in the world containing the richest zinc ore deposit ever found. Only a handful have ever entered the gates that locked the rest of the world from the scientific wonder of fascinating minerals and mining history. Only those who ran the mine and those who worked the shafts that go over 2,800 feet below the surface know of the geological treasures that abound. On July 1, 1990, the gates will open to EVERYONE to learn the secrets and to see the breathtaking aura of fluorescent beauty created eons ago!

Opening Day Brochure
Sterling Hill Mining Museum

In 1986, thirty-two years after the closing of Franklin Mine, operations at Sterling Hill Mine came to an end. At this time the Borough of Ogdensburg put the mine site up for auction, and Bob and Dick Hauck saw the possibility of making a dream come true. Both brothers had been interested in mineralogy and mineral collecting for most of their lives, they were willing to put their life savings into the purchase and preservation of Sterling Hill, and were delighted when they succeeding in becoming the owners of the historic mine site in 1989. The Hauck brothers tried to interest the State of New Jersey in taking over the property as a historic site. When this effort failed, a committed group of volunteers formed the Sterling Hill Museum Foundation to develop and manage the undertaking. With assistance from a

Figure 101. The Sterling Hill Mining Museum and the change house and ore bins as viewed from Plant Street. (Photo: Courtesy of the Sterling Hill Mining Museum)

group of dedicated citizens, the foundation turned the abandoned mine into the Sterling Hill Mining Museum.

A tremendous amount of work went into transforming what had been the last working mine in New Jersey into an educational tribute to mining, an undertaking that required hundreds of hours of work, dozens of volunteers, and several years to accomplish. On July 1, 1990, the museum opened to the public and offered acres of indoor and outdoor exhibits, machinery displays, historical buildings, hands-on demonstrations, and access to many underground tunnels. In 1992, the Sterling Hill Mining Museum was designated a National Historic Site.

In September, 2000, the Thomas S. Warren Museum of Fluorescence was opened on the grounds of the Sterling Hill Mining Museum. The purpose of the Warren Museum is to reveal the beauty and potential uses of fluorescence.

Bernie Kozykowski was a trustee of the Sterling Hill Museum Foundation and was involved in the development of the museum from the beginning. Bernie speaks with pride when he says:

> I was vice president of the foundation. In fact, I got involved in that whole concept a long, long time ago. I still take credit for having brought Dick and Bob [Hauck] up [here]. I knew they'd have an interest. The question was whether

they were willing to pursue it. They did. The second time it [the Sterling Hill property] was bid upon, it was purchased by Dick and Bob Hauck. [The Haucks] put a half a million into it. They recognized that if they are going to hold it privately, they are going to be paying taxes on it. The borough is still coming after them instead of cutting [them] a break. In reality, the Borough of Ogdensburg made it pure hell for Dick and Bob Hauck. One of the reasons that Dick kept as close to me as he could for the last two or three years was because he couldn't handle the Borough of Ogdensburg. I was the liaison with the borough. They respected me. The building official was ready to shut them down every time they turned around. I kept him at bay. [The relationship with the borough] is better than it has been. It's been getting better. Anyway, they [the Haucks] said, "Well, we're going to have to form a non-profit foundation." And I helped organize that with them, and became a component of that.

I started on the project in 1972, long before the mine company shut down, setting up the network that I needed to understand the issues. I gave part of my life to work for the company. It wasn't just to see where the minerals came from and put a paycheck on my table. I wanted to know about what was there because in 1972, a bunch of us recognized that that place was going to shut down. It was Gene Kline and myself and Steve Sanford, the three of us who originally talked it up — preserving Sterling Hill. Long before anybody else.

[In the work], I invested volunteer time. Expenses in preserving the place. Doing a lot of traveling in the name of the project. Telephone calls like you wouldn't believe. Time that came out of my practice that could have been used for income purposes. If you take all of that and you add it up over a period of almost ten years, it comes out to over a quarter of a million dollars by the time it's done. But, it was my choice, and I don't regret that. I would do it ten times over. Principally because the project is so much greater than anything any one of us might think of. It's worth every bit of that.

One of the things I'd like to say is that no matter how I might criticize it, at the same time, it's still better than anything else anywhere. I think it will last forever, and I think it should. I'd like to think that in large part, I helped preserve the place.

Figure 102. Narrator Harry Romaine and Mrs. Romaine. (Photo: Carrie Papa)

Like Bernie, many of the other narrators are delighted to have been connected with the Sterling Hill Mining Museum. Former miner, Harry Romaine, evidences this proprietary pride in discussing his first visit to the mining museum.

> I went up this one day and I met the man's [Dick Hauck] wife. So we were sitting there, getting ready for a tour, I was talking to her, telling her how I worked in that mine. She says, "They're going on a tour pretty soon. Come on, I'm inviting you to go along." And her and I walked in the back. And he was good. He has an education of mining. So then we come to the one place and he says, "This is the powder magazine, this notch in the wall," he said, "This is where they kept their powder." And I says to his wife, I said, "They didn't really keep the powder there," and I told her about it. She says, "Really?" I said, "Oh, yes," I said, "I took care of the magazine, five hundred foot level." But he knew a lot that he had picked up.
>
> Then we went from there down to what they called the change house. Well, my basket still hangs up there! Some of them were taken out, but where I changed, that was mine all the way through. That was hanging up there yet! So, after the people went out the far door, she got her husband to come over and we was talking. Before I went out of there, he wanted me to take a job as a guide. He was a nice fellow. He has a lot of good ideas.

~

Steve Misiur was a vital part of the Sterling Hill Mining Museum project almost from the beginning. After years of mineral collecting on the Buckwheat Dump, joining the Franklin/ Ogdensburg Mineralogical Society, and making friends with Dick and Bob Hauck, Steve also became a part of the mining museum endeavor. After the Haucks bought the place in 1989, Dick Hauck suggested that Steve work with the Canadian firm that was removing a rock crusher from the eleven hundred level. Steve recounts that, after successfully helping to remove the crusher, he was offered a job working with John Kolic, who was blasting out the Rainbow Tunnel to make that a part of the museum complex. After that he continued with other jobs at the mining museum. As Steve emphasizes:

> It was nice for me to be able to go underground here at the place that I tried to work in as a miner. All that history and all the wonderful things I've seen come out of here. Just being here was a privilege itself. The tunnel that we worked on is called the Dynamite Room. It's opposite of what is called the Rainbow Room in the mine. We did that one in thirty-five days, 119 feet long. The other tunnel we just opened up — the Edison Tunnel — John and I blasted out that new tunnel, 6,335 feet, in two years and twenty-one days.
>
> We've since added more tunnels. We want to be able to open up [another new section] sometime next year. That's our intention. You see, I adopted the attitude that Dick and Bob have about museums and places of attraction of this sort

Figure 103. Narrators Richard and Robert Hauck, founders of the Sterling Hill Mining Museum. (Photo: Carrie Papa)

— always have something new, keep the momentum of newness involved. When I was a young boy, that pebble pup, I did get to go to the American Museum of Natural History in New York to see their mineral hall. At that time, it was the J. P. Morgan Gems and Mineral Hall. It was an old-fashioned hall, and after visiting that the eighteenth or twentieth time, it got a little stale, with nothing ever changed. You can go there years afterwards and it's déjà vu all over again. There's nothing new. Our principal here is always add something new. Not every year, but enough that people will constantly be renewing their sense of wonder about the place.

I'm here on a part-time basis, right now. I have to juggle several hats. In addition to being curator and tour guide, I'm the editor of the newsletter. I see a grand future, but I think it will take still a number of years. It's probably going to be this year [1995], five years after we opened up, that we are going to finally break even in terms of the attendance we need to be able to have the income just to pay the bills that we have. In other words, people say, "Well, you're probably making a hundred thousand dollars a year." I say, "Yes, but it's costing us a hundred and one thousand to run it." But we think we're going to break even this year.

I'm just giving a hypothetical sense here. But [in the future] I see a place where the mill is going to be built up. It's going to be where people are going to go down into the mill — there's three huge rooms down there. We think we're going to be able to open them up to an auditorium. You'll sit down for ten or fifteen minutes, look at a film, a video presentation of what the mine is about. Then go from there into the mine. Walk on to all the different places.

Then, hopefully, we'll be able to go up to the top of the hill and get to see the building where the processing of the ore took place. We want to incorporate that to show you the production part. What happened to that ore after it comes up to the surface? That's the missing link. We want to show people how it was taken out of the ground, how it was pulverized, how it was separated before it was shipped from here to the smelter in Palmerton, Pennsylvania. We want to be able to show people how this resource will be made into zinc metal and then from there into your daily life. That's what we want to show. More of that cycle, from resource to man's natural use of it. I see that future here.

I'm on the board of the Franklin Museum as well. There's so much overlapping, it's in my vested interest to see all of these places work together and complement each other. For instance, the Franklin museum has a wonderful fluorescent exhibit. Here, we have the actual mine that you can go into and see where these minerals came from. The Franklin/ Ogdensburg Mineralogical Society has this specialized interest in furthering the knowledge of this area and they also conduct field trips in the area for you to be able to collect minerals.

There is so much overlapping that there is no sense of competition. There is a sense of pride for what each institution does. The mineral show itself is the grand finale for all of us to be able to bring people here and see this place.

Former Sterling Hill geologist Bob Svecz, who did the surveying for the Edison Tunnel, now serves as a tour guide at the mine. If there is any free time while he is at the site, Bob works on cataloging schematic diagrams and maps that Rutgers University has returned to the mining museum. Here Bob speaks of his work surveying the Edison Tunnel, the cataloging process, and the future of the museum complex.

> When I did the surveying here for the new tunnel and the Edison tunnel, I did it along the old New Jersey Zinc Company coordinates. I did not use the land survey marks that they have around here. With the maps that were still here, we could tie in exactly what we have here now with what was here underground. My desk used to sit right over there, and that is directly above the east vein strip on the three-forty level and the thirteen-forty north section that runs right through it.
>
> When this mine closed, the maps and sections were sent down to the Bureau of Mines, to be held on file for future reference, so that you're not going to build something on top of a mine shaft. When they put in Route 15 down by Woodport, they were putting in the pilings for one of the supports for the overpass. They could pour concrete into that hole and never fill it because they drilled into a main incline for one of the larger iron mines in the state of New Jersey and it wasn't on any of their maps. They were building a highway right over the top of it.
>
> These schematic diagrams were returned to us from Rutgers. I'm supposed to go through them and catalog them.

A Mile Deep and Black as Pitch

Of course the card catalog didn't come along with them. If you want to get the specifics on anything that was here, and a lot of things that were at the Franklin Mine, like the bell system for the main shaft, the Palmer Shaft, the schematics, the wiring diagrams, they are still on file here. There is so much detail, they even have the schematics of our latrines. The dimensions of it. If you wanted to know where the old leach field was up at the Franklin Hospital, that's here. Old railroad maps. If you wanted to know where the icehouses were in Franklin, they're located on property maps. Water maps. A lot of other things. They even indicate when a house was torn down or if it was destroyed by fire, there's a date on it. The land grants. The deeds. I'm hoping to get it all together with the catalog.

Now the question is whether I have the time, as well as giving the tour. They're in bad need of tour guides. Today, we get a bunch of school groups in and yesterday was a mad house in here in the afternoon. We needed five tour guides at one o'clock. That's how many tours we had going all at the same time. So I'm busy with that and I haven't gotten down there [into the archives].

It's [the museum] come along better than I would have thought. Seeing what you had to deal with, there was the fear that it's going to end up just like the mill yard down in Franklin. I mean, it's a blessing to see that it's come the way it has. Right now we have access to the mill on top of the hill, but there are no funds for it. It will cost quite a bit, but the mill in a sense is still operational except there's no power coming into it. The transformers that they had to supply the power for it have been taken out. They would have to be brought in to operate the machinery that is up there, the motors and everything that would run the conveyor belts. To have this operational someday is a goal.

My interest here is in the historical part of mining because it's not only a part of history, but it was also a part of my life. I can actually come back here. I've actually experienced it. I'm a living part of history.

Several former miners at Sterling Hill now serve as tour guides. They concur with Bob Svecz that they enjoy going back into the mine as guides. With real pride former miner Clarence Case exclaims:

The effect mining had on my life, I can go back today and be a tour guide. It reflects back to the days that I worked in the mine. And I enjoyed every minute of it. I can go back and do the same thing that I did years ago. Tell the people of my experience. What I did. How I did it. How I blasted. How I drilled in the mine. I love that. I'm happy to have the mine facility here. Absolutely I am!

I think the Hauck brothers are doing a tremendous job. If it weren't for those people, the mine wouldn't be what it is today. It may have been history, period. I very much like being involved with the museum. Absolutely! I wouldn't have it any other way. It's open seven days a week, with the exception of Thanksgiving and Christmas Day. I try and put in as much time as possible between that and my job. I try to get down every day for a couple hours, at least. I've still got my tag to show that I worked in there. I loved it.

After the Hauck brothers acquired the Sterling Hill property, they tried to persuade the state to take over the site, then tried to find investors in the Sterling Hill Mining Museum Project and when that failed, finally decided to take action themselves.

Sterling Hill Mine was at the edge of something happening. A number of people tried to get state influence to be able to preserve this particular place. But, apparently, either moneys or courage were lacking. Then we tried very hard to get a number of wealthy, influential people to be involved. Our dream was to have the Williamsburg of mining here in northern New Jersey. We have the facilities worthy of that attention. The Rockefellers, of course, did Williamsburg. We have Sturbridge. There's many of these preservations of these particular time periods or cultures or whatever. Unfortunately, many people who could have gotten involved had the good sense to realize it was not practical, not possible, although it might be [historically] worth it. So very few people that we saw had the courage or the insanity to get involved.

But, after just selling a valuable piece of property in Bloomfield, we had what we thought to be a substantial sum of money. We thought it was enough to save the place. We didn't know any better, so we did it. We had tried so hard to get others to take over, it was felt we should put our money where our mouth was and do it, so we did. And I have to include our wives' input, and our children's sacrifices, because we're spending our children's inheritance now. As we like to say, if they're any good, they don't need it; if they ain't any good, they don't deserve it. In either case, they ain't going to get it. [laughter]

Both wives intelligently looked at this project and realized its many, many pitfalls. Concerns about the liability insanity today, because if somebody drops a coffee in their lap, obviously somebody else is responsible. So, in this day and age, if you own property, you automatically take a tremendous liability. You can insure yourself to death, but it doesn't cover everything. So there's that concern. This other concern's about the usurping of our future. Can we make the commitment without totally surrendering our own ability to do things that we want to do once in awhile? All these sacrifices were looked at and the ultimate decision was let's give it a shot. Wives are both involved. Children peripheral, not full-time. Part-time and when they can. We, to this day, of course, do not regret this decision. That's the acid test.

But looking at it from a dollars and cents point of view, for the hours that are put in, the money that is invested, it has to be the worst investment we ever made in our life, monetarily. But, as they have said so often, they don't put pockets in a shroud. Putting a lot of money together so that your children fight over it, or the government gets it, we're not really interested in that. If we had a choice of getting all our money back, but the project failing or losing every nickel we ever put into it and the project outliving us, we would accept the second. That's our attitude. So we've made some significant and serious financial sacrifices to have this happen. Robert is salaried, but he is making less than he would if he was a parking-meter maid. And there's still not enough money in the organization to pay me a salary after five years. I'm hoping that someday we'll be successful enough that we'll be able to pay not only our employees a better wage, but management a living wage.

The only way we are able to meet our bills is we had to refinance the mortgage. We had to spread it out further. Our gross budget is around two hundred thousand dollars. But you take utilities are a thousand dollars a month between electric and heat. You take insurance bills, between liability, workmen's compensation, and other types of insurance;

we're somewhat shy of fifty thousand dollars [a year]. So you can see how the dollars are very easily taken up. We have twenty-five thousand visitors a year now, which is quite commendable in a short period of time. But we have to reach forty thousand people per year to make this place truly self-sufficient, where we're not robbing Peter to pay Paul.

The successes are many. One of the most pleasurable results of this involvement was meeting and making friends with some of the most unique and fantastic people on this planet. There's a lot of very important, great people we've run into that are participating in this project. One thing that is helping us that we're very proud of is simply the situation of having outside corporations recognize our value. They have helped us on some of our expansion programs. For example, when we did the Edison Tunnel, the Edison Fund board of directors was willing to give us a substantial grant to pay a major portion of that particular expansion.

We tried for one government grant and we learned a very important lesson. Lesson number one, never ask for a small amount of money. We asked for twenty thousand dollars. We felt if we asked for a little bit, we would increase our chances of acceptability. You know, "give them a couple bucks." When you ask for a little bit of money, they disrespect you. It's not worth their trouble and, obviously, you're insignificant.

The second mistake was to assume that you are judged on merits, not by, let's say, lobbying. When I mentioned to our local senator that we never got a grant up here after we tried very hard, the senator said, "Well, why didn't you tell me you were applying for a grant. I could have pushed it along a little for you." Well, since everybody else obviously knew enough to have somebody pushing for them, and we had nobody pushing for us, guess where our grant went down in the pile?

So what we're doing now is we are going to the private sector. We are delighted to be worthy of a Geraldine R. Dodge grant for a program that we are running now for education. That Dodge grant was a challenge grant that we had to match in three months. We did it in two. To have earned the respect of the Geraldine R. Dodge people, you've got to be good. They recognized that we are responsible, prudent, careful. We spend our money very carefully here. We were able to match this through the Edison people coming back and helping us again, a Dorr grant, and we had several others that came in.

Fortunately, we have attracted two great people, Mikki Weiss and Alan Rein, who are very knowledgeable. Mikki Weiss is an educator. She's retired out of the New York school system and she knows the verbiage, the communication skills that are necessary to have a credible application potential. She was very successful with the Dodge people. Alan Rein is a businessman, who also is an educator serving on the board of education in his community. He also has been able to touch base, not only on a corporate level, but also on an education level. These two people have now added other people to their team, which has the potential of bringing us very successfully into the next century.

We have an expansion program at the top of the hill to preserve the milling end of our industrial complex here. We're going to need at least a half a million dollars to do that. To find the right doors to knock on, knowing how and when to knock on those doors, are well within the means of a number of people, who are finding the grants that are available out there. We're now, in cooperation with the Franklin Museum and Montclair State University, applying to the Dwight D. Eisenhower grant program.

There's something here for everybody. There are so many things. This place is the pinnacle of the universe in many people's opinions. Because of our unique mineralogy — you've heard the stories about more minerals found here than anywhere else on this planet, more minerals first described, more minerals unique, the best fluorescents on the planet. We've had historical activity here that goes back to before this was even a country. Tradition says the Dutch were here in the 1600s. We know Lord Stirling was here in the 1700s. And Sam Fowler, of course, succeeded in the 1800s in making this a commercial mining enterprise. A tremendous span of not only history, but commercial activities. The science of mineralogy, where in 1814, Archibald Bruce described the first American mineral, zincite, from this location. There's so much tie-in here that goes along with the fabric of how this country grew. The trials and tribulations the country went through, the recessions, depressions, the panics, they're all written in the history of this area. Even the sad history of a proud old company, the New Jersey Zinc Company, having a hostile take-over by Gulf and Western. Stories to be told.

There's such a diversification here that, maybe you're not into minerals, but maybe you're into machinery. Maybe you don't like machinery. Maybe you like history. Maybe you don't like history, there's still something here for you. Our most delightful tours are when we have one of our real miners give a tour. Our tours are an hour and a half to two hours, but we could keep you busy here for four hours. We have put in over a thousand feet of new tunneling. Our exhibits are greatly enhanced. The museum has come alive. It's got a whole charisma to it. The little kids walk in and go like, "Wow!"

The Smithsonian exhibition gives us the credibility and reputability we need. We make the claim that we are world famous, unique, special. Certainly being one of four featured exhibits at the Smithsonian doesn't discourage that type of claim. They've had staff meetings up here three times where they had as many as twelve people on the property examining anything from rock texture to the mechanics of various situations.

We are very much up there with world class museums. I would say that we are in the top ten. We are here with purpose,

Figure 104. As Steve Misiur said, "When I walked into that adit . . . I felt like something said to me that I came home." (Photo: Courtesy of the Sterling Hill Mining Museum)

pride, and direction. The future, we've got all kinds of marvelous things we're going to try to aspire to. Eventually, this place will outlive us, which is what we want. The way it's set up with all the good people on the board of directors. As long as people care about science and history, this place will survive. That's our tie-in. With education, this place will be here a hundred years from now. Guaranteed.

In January 2000, the Carnegie Museum of Natural History in Pittsburgh, Pennsylvania awarded the Carnegie Mineralogical Award to the Sterling Hill Mining Museum. An award of international prestige, the Carnegie award honors outstanding contributions in the field of mineralogical preservation, conservation, and education.

In Our Nation's Capital

The Smithsonian Institution's Museum of Natural History in January will begin an $8 million face lift with the help of 6 tons of minerals and decades of mining expertise from the Sterling Hill Mining Museum in Ogdensburg, Sussex County, New Jersey. The Smithsonian will use the stone and mining equipment to renovate its gem and mineral hall and create what museum officials say will be the world's most comprehensive earth science display . . . . The Smithsonian will model its mine display on the 150-year-old Sterling Hill Mine, which will be one of four deposits represented in the exhibit.

Asbury Park Press — November 30, 1994

On September 20, 1997, the Smithsonian Institution's National Museum of Natural History opened the Janet Annenberg Hooker Hall of Geology, Gems, and Minerals. The hall is the setting for the most ambitious reconstructed environment undertaken by the Smithsonian Institution up to that time. Included in the new hall was a reconstructed mine based strongly on Sterling Hill Mine and a display of fluorescent minerals that includes many specimens from the Franklin and Sterling Hill mines.

In preparing for the new exhibit, mineralogists and exhibit designers from the Natural History Museum visited Franklin and Ogdensburg several times during a five-year period. After observing and studying the Sterling Hill Mine, the Smithsonian purchased six tons of minerals and fluorescent specimens from the mine to include in its mine reconstruction.

Figure 105. The legacy of the Franklin and Sterling Hill mines reaches out to the author's grandchildren as they visit the Sterling Hill Mine exhibit in the National Museum of Natural History's Mine Gallery. (Photo: Carrie Papa)

John Baum clarifies the connection between the local mines and the Smithsonian Institution.

> Oh, they've got thousands and thousands of Franklin minerals. They probably have nearly all of the 345 [different species], or whatever the current number is. It changes. They would have nearly every species that there is, just about. Most [first class science museums] have the common Franklin minerals. They all have fluorescent minerals, but in order to see the greatest variety and finest specimens, you'd have to go to the Smithsonian, Harvard, or the Franklin Mineral Museum. We have more than the rest of them presumably, but probably the finest specimens are at Harvard or the Smithsonian because they were collecting way back.
>
> I added to my collection of Franklin minerals [when I was working in the mine] and I've turned that over to the Smithsonian now. They've all gone to the Smithsonian. Although, I've kept one or two specimens of ore. For instance, I keep just pretty specimens that visitors to the house can

oooh and aaah over. I have a cabinet built into a family room there, which is concealed. I don't know why we bothered to conceal it, because I insist that everybody see it that comes to my house. [laughter]

The miners, in the old days, used to take a broken franklinite of some size and repair it with plaster and use stove blacking on it and sell it as an intact specimen. One of the best specimens of a Franklin mineral down in the Smithsonian turns out to have been repaired. Worse than that, pieces from this species were all hooked on and then the glue was tastefully covered. For years and years and years, that [deceptive piece] was featured in articles and turns out to be a fake mineral, so to speak.

They are setting up a room or a cave or stope, if you will, a working place [exhibit in the Smithsonian] that will be fluorescent. If I'm in Washington, I will [go see it].

Several other narrators mentioned particular Franklin collections that were in the Smithsonian. Many also reported on visiting Washington for the sole purpose of seeing the mineral specimens from the Franklin-Sterling Hill area that were on exhibit in the national museum. For example, Ann Trofimuk speaks of her brother Nick's mineral collection.

Now, Nick — again, I'm not prejudiced — but he was very intelligent. The proof of it is the direction that his life took in the mine. He went to work in the mine, and, I think, if you get into minerals, you will find that he is one of the best sources of minerals and data on minerals because he opened his mind to everything that was there. He worked with a geologist. He learned many things about ore bodies. In fact, so much so that the head of the Mineralogy Department at the Smithsonian — I forget what his name is, doctor something or other — came and talked to Nick a number of times to get information on the ore bodies and what he found, and where things were, and the minerals.

Pauline and Nick went on a trip to Washington and this particular doctor, whatever his name is, said if ever they were there to be sure and let him know. He took them behind the scenes where they had to go through all these security measures, the fancy buttons to push, and names to give, and everything. He took and showed them some of Nick's minerals that are there at the Smithsonian. He said that there probably would be one that would be named for him. That nobody else had discovered it.

Part of Nick's collection he sold to Harvard University. There were a number of men that had collections, but not like my brother. He was really knowledgeable about those minerals. In fact, I think some of the geologists would come to him at times to ask him questions because he was working with it.

Former collector Ewald Gerstmann also mentions Nick Trofimuk's mineral collection:

Years ago, I bought a few [specimens from Nick Trofimuk]. Harvard and the Smithsonian bought his collection. They have [Franklin minerals] in the Smithsonian. I used to go to the Smithsonian, looking at Franklin minerals. To see what the museum had. They had quite a few behind the scenes.

I've been behind the scenes thirty years ago. In fact, the curator of the Smithsonian, his name is right up there on the honor's dedication. [framed certificate on the wall] He wrote a book on me. Paul Fazeltof — I can't spell that. He's dead. It's on there — Award of Merit, Franklin Mineral Exhibit.

Laura Falcone remembers that she was in grade school when she was told about Franklin minerals being in the Smithsonian Institution.

In geography [class], one of my old school teachers, Mrs. Higgins, I think, told the class about zinc mining. We were always taught about it. That Franklin was one of the most wonderful mines in the world. That Franklin was the model mining town of the east coast. And that our minerals were in our nation's capital. In the Smithsonian. At that time, I had no idea what the Smithsonian was, but I know I felt proud about it — that our minerals were there. Of course, we knew that Washington, DC was the capital of our country. And I remember it just seemed very wonderful that Franklin minerals were part of our nation's capital.

Mrs. Higgins always said *our* minerals, as if they belonged to us personally. Actually, I don't remember Daddy ever bringing home any rocks or stones. But it didn't matter if we actually had ore in the house or not, we still felt the pride of ownership in having our minerals in the Smithsonian. It was a sense of community pride, of belonging, of being a part of this very special town of Franklin. This very special place that had its minerals in the Smithsonian.

And it wasn't only one teacher that developed this sense of community pride in us. All of the teachers seemed to tell us about Franklin. I know I heard about our minerals being in Washington many times. I just read someplace that the Smithsonian is doing some kind of mine replica and that Sterling Hill is part of it. I hope I can get down there and visit it. I'm still proud of Franklin and *our* minerals.

Linda Deck, who was among the Smithsonian's visitors to Sterling Hill, has been associated with the new exhibit project since 1990. In her interview, Linda explains many of the details of the extraordinary ten million dollar undertaking.

> The first inklings of working on a new exhibit came in the late 1980s. They started thinking about it. The previous gem hall was opened in 1958, so it was thirty-six years old, it was getting on in terms of permanent exhibit halls. [In 1995] we closed our entire geology hall complex, which included gems, minerals, meteorites, geology, the moon rocks. We closed that

Figure 106. Narrator Linda Deck. "In our Mine Gallery, we re-create four specific mine localities in the United States that help to illustrate where minerals come from." (Photo: Carrie Papa)

entire space so that we could do a very coordinated renovation of the entire thing and really knit it all together.

One of the reasons that it was argued that we should do a new hall is that we had acquired so many wonderful specimens over those thirty-six years. Because of size, a particular specimen wouldn't fit in the place that would be appropriate in the old exhibit. We wanted to put out our latest and greatest and best specimens.

The planning team consisted, I would say, of about two dozen dedicated people. The production team is perhaps, oh gosh, if you take everybody into account that's working on it, a hundred or more. And everyone has his own ideas, his own agenda, and [it takes cooperation] to get that all firmed up and moving in one vector toward a good end point. Just keeping everybody focused on that end point, making those negotiations and coming up with the best product meant arguments. Of course! Daily! Hourly! Oh, of course. People have different ideas about what is the most important thing to say, how to say it.

Here at the National Museum of Natural History, I am project manager and exhibit developer for our permanent geology and paleontology exhibits. So, right now, I'm working on this Geology, Gems, and Minerals Exhibit. Dr. Jeffrey Post is the lead curator for the exhibit. He's a mineralogist, so he's the one that oversees everything that goes on in terms of the curatorial and content part. It really is a major effort, taking much time and ten to twelve million dollars.

In our Mine Gallery, we re-create four specific mine localities in the United States that help to illustrate where minerals come from, the diverse geological environments from which minerals come. We've always planned four because we wanted to get the diversity in there. One of those environments is a pegmatite deposit. That comes from Virginia. Not very far from here. It has topaz and other rare minerals like that, but the biggest, the most rock that they mined is called amazonite. It's a turquoise cold rock — it's not turquoise now — but it's turquoise cold rock that they use for lapidary work. So that's the major product out of that mine.

We also feature Bisbee, Arizona, which is a copper porphyry deposit. We feature Sterling Hill, which is this really interesting fluorescent zinc and iron deposit, and we feature what's called a Mississippi Valley Deposit, which is also a lead and zinc enriched deposit but very different from Sterling

Hill, which is unique. It is unique in the world. What an incredible deposit, especially with this fluorescence feature that it has. Just unique.

We selected the four [mines] and then we went to those four specifically. The reason the four were chosen is because they're so different, so I really enjoyed each for a different reason. Sterling Hill was fabulous. I tell people, "Gee, are you going to be up in New Jersey? You really, really must stop there. It's a great place." I loved the mine itself, especially with the fluorescent display. I really liked that area that showed the miner's clothes and equipment hanging up, all of their things because it really allows you just as a person to put yourself in the place of what a miner's life was like during that time.

That's something that we do try and do, to make that connection. Not just — here's the facts, here's the natural history, here's the geology. But what does this mean to you as a person? How does this relate to your life? Could you see yourself as a miner, for example? I think that's nicely done at Sterling Hill. What the Mine Gallery does is, it actually creates these vignettes so that when you walk into the Gallery and turn a corner, there in front of you, you're looking down a tunnel in the pegmatite deposit. You walk a little farther and you're looking down a tunnel in the porphyry deposit. We're trying to re-create what it looks like at that mine. We're using simulated rock to create the rest of the mine, the matrix of the mine, so that that would hold it all together and create that experiential feel that you're entering a mine, and it's carved out of the earth.

We also will have a video. The working title of the video is *Ore to Product*, where we show, through video clips edited together, the process from taking the ore out of the mine, then processing it into some product. All of the sounds and sights associated in the process. This is from an actual working mine.

It's a national audience that we get in that hall, so we're writing it at an educated eighth-grade level. Especially families make up a great portion of our visiting public, so we didn't want to pitch it just to the buff [devotee], you know. That would be too pedantic. Or certainly not to the professional. That's not who it's for. It's for the general public, for our visiting public. It's not a children's hall either. I think it will be fine for everyone. We will have some complex concepts in there, but we try to explain them very well.

Our mission here is that we are dedicated to understanding the natural world and our place in it so that guides us. That mission statement is behind every effort here of the entire museum. But, as far as the mine exhibit itself, we have major messages of understanding — where minerals come from, that they are building blocks, that minerals matter in your daily life. We are the ones that show you your world around you and help to interpret that for you. Help to bring it closer to you. Help you to see it in a different perspective that you never thought of before. Like these minerals that, here they are, they're beautiful objects, beryllium for instance, or lithium in a beautiful crystal, but did you know we use it for drugs and that sort of thing. Appreciating it not only on the aesthetic basis, beryl, beautiful beryl crystal, but then appreciating that beryllium is used in making lighter and stronger alloys for the aerospace industry. Stuff like that.

The Smithsonian is, in a way, like a national park. It's a treasure that people want to come and visit. Every year, our visitation here is between five and seven million visitors. A specific new exhibit like this is a very big thing. And, this being one of the few major geology and mineralogy exhibits that will have opened in the country over the past few years, I think it will make a big splash.

I get a lot of satisfaction from my job and I think that's typical of people who work on an exhibit. It's providing a lot of satisfaction. It's something new every day. The best thing is knowing that what you've been working on will be seen by five to seven million visitors a year. A year! And you have the chance to really affect someone's life and give them a deeper appreciation for natural history. You know, you see the faces of the people. You know that as you plan something there and you get them to say, "Wow! I never knew that before." Or, "I never knew it looked like that." You know that's going to stick with them, and that is a wonderful thing about doing exhibits here.

It's really rewarding to work on a project that you start in its infancy and take it all the way through to an end point. Then you see a great product at the end. That's great. It's not just a file folder full of papers. It's a real thing that you can visit. People here really take pride in their work and enjoy the fact that their handiwork is going to be out there and a part of the National Museum of Natural History for so many years.

Just as the Smithsonian staff takes pride in their contributions in developing the mine re-creation at the National Museum of Natural History, those mining communities that have provided minerals for the exhibit take pride in their contributions. These mining towns, including Franklin and Ogdensburg, are proud that their local minerals were selected to be a permanent part of the national treasure that is the Smithsonian Institution.

Watching the Lights Come On

*The Sterling Hill Mining Museum is an extension of
the classroom designed as a three dimensional textbook
to show students the importance of mining in everyday
life.*

Sterling Hill Mining Museum Teachers Guide

The Mine Safety and Health Administration (MSHA) is concerned primarily with safety and health issues, but the agency also tries to promote an understanding of the work experiences of miners and the significant contributions that they and the mining industry make to the nation. MSHA posts information on its website, *www.MSHA.gov*, about mining museums, miners' memorials, historic landmarks, and actual mines that are dedicated to the study and preservation of mining cultures.

The number and variety of mining sites that are accessible to the public is quite remarkable. From searching for treasures at a plowed surface field at the Crater of Diamonds State Park in Arkansas to descending a thousand feet into a Colorado gold mine, there are exciting places to explore throughout the United States. One can, for example, visit an iron mine in Michigan, a copper mine in Utah, a lead mine in Missouri, a coal mine in Pennsylvania, and a zinc mine in New Jersey.

Mineral museums may display comprehensive collections of mineral species, and collecting sites offer an opportunity to dig up a rare crystal or unique specimen. Other museums feature exhibits that depict the daily lives of the immigrants who made much of the mining industry possible. Probably the most unforgettable sites are the memorials dedicated to those who lost their lives through terrible disasters in the mines of America.

The preservation and flourishing of all of these museums, memorials, and historic sites contribute to our understanding of the significant role played by the miners and the mining industry as they increased the wealth and industrial power of the nation. Through these various displays one can see a miner's life above and below ground, view broad mining and milling operations, and realize the significance of mining in day-to-day lives. As one Bureau of Mines publication states:

> There are few times, if any, that we stop to consider the importance of mining and minerals. Yet the quality of our life, our national security, the stability of domestic and world economies, science, industry, and the arts are all based on the minerals we mine from the earth.[1]

The need to improve the public's awareness of the importance of minerals is underscored by Linda Deck of the Smithsonian, as she states:

> That was one of the first visitor surveys we did. We just got out there and asked people what they think of minerals. And they said, "Well, you know, vitamins and minerals, we need that." Or we asked, "How do you interact with minerals on a daily basis?" Some of them were clever enough to realize, "Well, you know, I cook in pots and pans and those were aluminum, so I know . . . ." But [people were not aware] that they, virtually every minute of every waking hour and sleeping hour in fact, use minerals in some way. You know, the paint on their walls tinted with titanium to make it white and all that. They just didn't have that deeper appreciation, which is one of our missions for the exhibit.
>
> We certainly will be doing a lot of educational programming and interpretation in the exhibit halls. An interactive video in that area has the working title *How Many Ores?* How many ores does it take to make a car is what we're looking at. What we hope to get across is that not only are there so many more minerals than you would have thought that go into an automobile, but the volumes of ore

that you must start with to then get the amount of metal to use in the car.

We'll also have another example, "Okay, now we've shown you a complex one, how about a soda can? How many minerals does it take to make a soda can?" Of course you think of aluminum, but there's lots of other stuff that goes in there too, other resources as well. And this is an interactive [exhibit] so the visitor will have a chance to choose different answers, guess different things.

Bob Svecz, tour guide at the Sterling Hill Mining Museum, relates what he tells visitors about the importance of minerals.

When they come here they learn something about mining, which they didn't know existed, because they can't get it from any other place. You don't hear about mining in New Jersey anymore. You don't hear it in the schools. The kids in Ogdensburg, they can see us here, and they didn't know this was a mine. What we try to impart to them is how important mining is, that you cannot live your life — you never could live your life — without mining.

But it's not something you see. It's always in the background. If I ask somebody a question, "What industry is responsible for the manufacture of automobiles, what companies?" The answer I would get would be the automobile industry, or more specifically, Ford, Chrysler, General Motors. But, looking beyond that, if it wasn't for mining, there would be no automobiles. You can't name a piece of equipment or anything on that car that isn't directly related to mining, especially, the one thing, steel. You need the steel for the frame, the body of the car. The tires. Today, very few people realize it's well over 20 percent zinc oxide in the rubber of the tires, or else your tires wouldn't last as long as they do today. The plastic knobs on the radio are all derivative.

If you think about it — I use this for older groups of people — by the time I finish, I ask them, "What is the oldest profession in the world?" It isn't what you think. Now, think about it, it has to be mining. Pre-historic man, cave man, he used stone knives, axes, arrows, or whatever as his tools. Where did he get them from? He got them from the earth. That's mining.

A Mile Deep and Black as Pitch

Steve Misiur, in addition to being a tour guide at the Sterling Hill Mining Museum, also serves as curator, newsletter editor, and educator. When asked what he liked best about his work, Steve replies:

My most directly rewarding experience is taking [visitors on] the tours and watching the lights come on. Particularly when I have a group of people who come here, who want to be here, and who want to know what's happening. We have an expression we use here — that the lights come on. We don't mean the lights in the tunnel. We mean the lights that come on between the ears. Getting to see that reaction in people, that's probably the most direct sense of reward I get out of this place — taking a tour group through and watching them.

I had a tour about four years ago. There was a young boy on my tour who was a little wiseacre. He knew the answer to almost everything, but not quite the right answer. I was intrigued by this little boy. Then, out of curiosity, I asked him at the end of the tour, I said, "Excuse me, young man," I said, "where does water come from?" He shrugged his shoulders, "I just go to the kitchen, turn on the faucet, there it is." He had no conception of what happens before that. That's what we're trying to do here. We're trying to show people, this is how it starts from here before it gets to your house. That's what our education here is about. That's what those grants are going to be used for, to try to get people to think before time, to try to show them the natural world, what we use out of that natural world for our daily living. It's very important because people have lost that sense of connection.

I find personalizing my experiences with the tour groups gives them a special feeling that this is the actual place and we have done work here. They get that sense of realism that way. We do nothing fake here at all. We tell it like it is whenever we can. We have to do that, especially with school kids. School kids are very sharp. They pick up on phoniness real quick so I do nothing phony with the kids. I talk at the same level with the kids. I treat them as adults because I know they have intelligence. They may not have the sense of appreciation. Maybe that's part of what I'm trying to do here, convey that enthusiasm to these kids. Try to draw them into the sense of wonder that I have about the place.

That's the one chief thing that we want to do with our educational program. We want to be able to find or to set up an environment where you can convey as much meaningful information as possible in the time that people are here. But we have to be mindful of the saturation point of people. We find, from our experience, that two hours is the practical limit that you can inundate somebody with all of this information. Because it's an alien environment. Most people who come to this mine have absolutely no idea what a mine is like. All they've ever seen about a mine is maybe on these old fashioned B-grade movies that are watched at midnight.

Probably the other rewarding part of the job is to try to find the right balance of conveying the information while keeping their interest up. And on top of that, the triple duty is to make it enjoyable. See, people have a very bad feeling when you use the word educational. Like, oh, you're sitting in a classroom listening to a talking teacher. So we want something here that people would want to come back to. Our grants are all for educational development. The Rock Discovery Center, visiting school groups will be given a little descriptive sheet and they'll get a little box made out of compartments, and they'll be asked to pick out the rocks and fill their box of minerals. So they'll kind of go wandering around and if they look at coal, they will see that it's black, real glassy, soft, and light, and they'll put it in the respective compartment in the box. Or a piece of slate, or marble, or sandstone. So they'll be given this descriptive sheet, a little box, wander around the Rock Discovery Center, and hopefully fill up that box. We want to give them this type of hands on experience. What makes a lasting impression on most people is that they are able to actually find it themselves. When they think about what they're looking at, and then, "Hmmm, I think this is it," they have a greater appreciation for it. What the kids find, they keep. It's a challenge to them. Making you think, getting you off square one, that's the whole thing.

Founders of the Sterling Hill Mining Museum Dick and Bob Hauck speak of the new Rock Discovery Center and their educational goals for the museum. Their thoughts and words are combined in the following narration.

We're presently anticipating a grant from the Merck Corporation to help us with a children's collecting area, called the Rock Discovery Center. You have to be careful how you do that. You can have a pile of rocks over there and say, "Kids, go knock yourself out, pick out rocks," but that's not enough. What we're developing is a geology collecting site where the kids will actually learn elemental geology. That's what the whole purpose of it is. It's more than just pick up some pretty rocks, put them in a box, and go home. I mean, that's not really doing the school kids a heck of a lot of good. They have to identify what they take or they don't go home with it. We have twenty-five thousand visitors a year now, which is quite commendable in a short period of time. Sixty percent school children. Mostly coming from within fifty miles of our establishment. But there are areas we haven't even touched upon. We're at the top of the state, so we are closer to many places in Pennsylvania and New York State than we are to many places in central and south Jersey. The horizons that haven't even been touched yet are limitless.

We're now, at the present time, in cooperation with the Franklin Museum and Montclair State University, applying to the Dwight D. Eisenhower grant program. The mineral show helps us because it brings people an interest in the area. It was an excellent show this weekend that just passed [October 1995]. They had a new gimmick in making the show for three days, where the school kids came in on Friday to see the show. Most of those kids came home, "Mommy, Daddy, let's go to the mineral show." It was a really good tie-in. So it's growing. It's building.

The future, we've got all kinds of marvelous things we're going to try to aspire to. We are very appreciative of the mineral collecting community, the 2 percent of the world that they are. We find tourism to be a very satisfying and very important support factor, as the 25 percent that they are. But the 60 to 70 percent that is school group oriented — we'll throw in scouts and other groups in a similar situation — there's where our strength is. There's where our goals are. There's where our future is.

The most complete report on the educational programs at the Sterling Hill Mining Museum came from Susan Cooper, a fifth grade teacher in the Ogdensburg Elementary School. Susan developed a Sterling Hill Mining Museum curriculum for her school and spoke of her involvement with GEMS, the Geological

Figure 107. Narrator Susan Cooper was very active in developing the educational programs of the Sterling Hill Mining Museum. (Photo: Carrie Papa)

Environmental Mineralogical Sciences education project at the museum. Susan's commitment to her students and her interest in preserving their heritage is visible throughout the interview.

When I first started teaching here, I started getting my students involved with the history of their hometown. At that time, we learned that the mine was closed down and the area was about to be sold. When the Haucks were looking into buying the mine, my students made a presentation to the community members telling them how important it was to preserve the heritage of the town, and to make sure that everything that their grandparents and great-grandparents had done was still there and could be studied. We met the Haucks and we just became good friends, and it developed from there. They did end up being able to purchase the property, and turning it into a museum. Since then, I've been bringing my own students down to the museum.

Through the museum, I was invited to come down there as an educator and develop a teacher's guide for other teachers to come and visit. This was part of the GEMS project. We had a major grant, fifty thousand dollar grant from the Geraldine Dodge Foundation [to develop the teacher's guide].

The GEMS project training lasted two weeks for me. This was done by the original GEMS team, headed up by Mikki Weiss, who's the educational consultant down at the mine.

In October, I'll be teaching a workshop for teachers throughout Sussex County. Last summer we started the Statewide Systemic Initiative, where we trained teachers from Pennsylvania and New York during the summer months. We're training them about coming here, but also we're training them in geology and mineralogy and the mining sciences just to make them more knowledgeable so that they can transfer this [to their students]. The GEMS project is ongoing. We're still continuing to outreach to other people throughout New York and Pennsylvania, and in the future plan on being on the internet and spreading out even further.

The teacher's guide is totally separate from the curriculum, which is something that I wrote just for my school. That's a smaller endeavor. After I became involved with the GEMS project, I learned more and more about the museum so I decided — knowing so much about mining and minerals — that I would write a curriculum for the school that would allow all the different levels to go down to the museum and not do the same thing.

So I set up a different type of activity for each grade level. One grade level would learn about the mining history. Another would learn about simple machines and how they are used. Another would learn about environmental concerns. Throughout all of the activities, I weave a lot of different subject areas; the language arts, of course, were easy to fit into anything. We do a lot of writing and we do a lot of communicating with members of the community and that type of thing. Science is extremely easy with the rocks and minerals. And social studies, of course, and the history of the town.

I think it's very important to make sure that when children are learning, that they learn a lot of different subject areas at one time, rather than placing importance on one. I like to teach science and math together, and at the same time, have them learning their language arts skills. Things are integrated. That's one of the major thrusts [of the curriculum].

I have my fifth graders acting as mentors for the younger students. For example, for the second grade, we go down to their classroom and do a pre-activity with them a week or so before our trip to the mine. Then we take them. The fifth

Figure 108. The miners' change house of the Sterling Hill Mining Museum now displays mining artifacts and mineral specimens. (Photo: Courtesy of the Sterling Hill Mineral Museum)

graders would be paired one on one with the younger students and they'd bring them down to the mine. And they'd talk to them as we walk down about what we're going to see and everything. It's not just your straight academic subject areas. It's a great growth in being able to communicate, to help one another, to feel good about yourself, and that type of thing.

For the other classes, I have a setup which would last maybe two weeks, a pre-activity, then a trip to the mine, and then a post-activity a week later. Then they continue on up to eighth grade. But since I'm the one that's coordinating the entire program, my fifth grade students have more interaction in it, more exposure to the whole deal. But every grade level gets about a two-week exposure throughout the school.

The museum is just phenomenal. Every time I go, and I've been there hundreds of times now, but every time I go, I see something new and different. It's still fun for me and I try to relate that excitement and that fun to the students. Most recently added is a Rock Discovery Center. It's piles and piles of rock that the kids first have to know about and then have to go and find. They'll learn the different characteristics, the texture, the cleavage, and all of that. And that's going to be developed even further. We'll have more of a scientific-type shed there with beam balances and a lot of different equipment for identifying the rocks.

The number of visitors at the museum had reached over thirty thousand by the year 2000. There is every reason to concur with Susan's assessment of the destiny of the Sterling Hill Mining Museum when she says:

> I see the future of the museum as an educational facility that's just going to continue to grow and grow.

1. US Bureau of Mines, *This is Mining*, 30.

The Enduring Legacy

It is a pleasure to attend the grand opening celebration of the Franklin Heritage Museum at a building that has played an important part in the development of Franklin during the first half of the present century. The location of Mill #2, the Palmer Shaft, and these associated buildings at this spot marked the rebirth of Franklin, and these buildings in a sense are the Ellis Island of the local zinc industry. Passing through these doors for the first time we were initiated into the fraternity of the Franklin mill, mine and surface gangs — men whose very lives depended on the skill and dedication of their fellows. That the mine is finished now is the nature of the mining business, but that its memory is to be retained in this splendid museum is thanks to dedicated individuals interested not only in the mine but also in the sociological impact which it had on the community.

Opening Day Remarks,
Franklin Heritage Museum

As has been noted previously, mining is essential to a society's needs. Minerals are necessary not only to meet basic requirements, but also to achieve and maintain a high standard of living. It has been estimated that each American uses about forty-seven thousand pounds of mined materials each year. The world population of humans is expected to reach seven billion by 2013, and since developing countries are struggling to improve their standards of living, the demand for minerals will grow.

While known reserves of natural resources are being consumed at an ever-increasing rate, it also is true, contrary to popular belief, that neither the United States nor the world is running out of minerals. Although high-grade deposits have become more difficult to find, advances in technology mean that lower grade ores will become economically feasible to mine. For instance, at its Carlin property in Nevada, Newmont Gold Company is mining "invisible gold" that makes it possible not only to increase the amount of gold mined each year, but also to build additional reserves. Other mining companies are accomplishing similar wonders.

America's leading mine operators — companies like Phelps Dodge and Kennecott Copper — continue to play a tremendous role in the nation's economic growth and prosperity. They also help to support a stable lifestyle for their employees and the other residents of their mining towns. The management and the employees of such companies would agree with Newmont Gold's general superintendent Tom Enos when he says, "We're not just a bunch of miners who've come in here and pitched our tents . . . . We're committed to the school districts, the communities and the housing."[1]

At one time, the employees of the New Jersey Zinc Company felt much the same way about their company and the communities of Franklin and Ogdensburg. The mine and the company provided a way of life that is fondly remembered today. When the New Jersey Zinc Company closed the Franklin Mine in 1954, change in the community was inevitable. However, contrary to the fears of many, the town did not just disappear. The residents already had most of the services, such as water, electricity, schools, and a hospital, library, and community center, that would be necessary for continued viability and, possibly, growth and development. Once on their own, it was up to the people of Franklin to maintain and improve the facilities and services given to the town by the New Jersey Zinc Company. The town government, prominent citizens, and local residents had the challenge of making the future of Franklin worthy of its past.

Prior to 1995, there was no community group to promote Franklin's history. Nearly half a century after the closing of the

Franklin Mine, a group of interested residents considered forming an organization to encourage an interest in Franklin history.
At a meeting of Franklin's Economic Development Committee
the possibility of forming a historical society was discussed. Jimmy
Van Tassel, a lifelong resident of Franklin and an employee of
the borough, attended that particular meeting and became one
of the moving forces behind the founding of the Franklin Historical Society.

As a part of both Franklin's past and present, state senator
Robert Littell addresses the need for people to be involved in
their community.

> You know everybody was concerned when the Zinc
> Company closed down. They thought it was the end of the
> town, that it would become a ghost town. It never happened.
> The prospect of Franklin's future is bright. In 1964, we set
> out to make Franklin a town which had sewer lines running
> through the town connected to a wastewater treatment
> facility. The town has been aggressive in upgrading the water
> supply, putting in a new tank, and new lines, and fire
> hydrants to make sure that the volume and pressure is good,
> to make sure that there's fire protection. The fact that we
> have sewage available in town is going to make it more
> attractive than other communities that don't have sewage
> available to them. Because it's going to be very difficult for
> somebody to come in and locate a factory, a business, a Wal-
> Mart, without that sewage allocation. When I was on the
> Franklin council in 1963 through 1965, we started that
> project. It didn't get finished until the 1970s, but it's there to
> provide appropriate treatment of the wastewater.
> Industry in Franklin is somewhat limited, but we have
> the industrial building at Munsonhurst. It used to be Hewitt
> Robbins. And it's now owned by Charlie Fletcher and his
> son. They've done a good job in fixing that factory building
> up and they've rented it out to a variety of small businesses.
> There's talk of a Wal-Mart coming to town. Weiss
> Supermarket is also a possibility.
> I am thrilled to see revitalization on Main Street and other
> areas of Franklin. Antique stores are a wonderful boost to
> economic development. My wife, daughter, and sister, and a
> few friends opened the first of a few antique shops, Edison

Antiques, in a building that my grandfather Watson bought from Thomas Edison when he closed down his mining operation up on the mountain in Ogdensburg and Sparta. He took that building apart, and brought it down here on a horse and wagon and put it back together. My dad used it when he was in business delivering and selling propane gas. My father pioneered the sale of propane gas in the community and we operated that business for many years.

I think the spirit of Franklin lives on. Many people have lived here and stayed here to make it better. We have an outstanding volunteer fire department. We have an excellent volunteer rescue squad. All of the components are here for a strong community. I spoke to a woman who had moved into Franklin recently and she was bothered by some young people harassing her, and she told me she wanted to move

Figure 109. Senator Robert Littell, his daughter Alison McHose, and his grandson Logan McHose. Senator Littell said that five generations of Littells have lived in Franklin. (Photo: Courtesy of Alison McHose)

out. And I said, "Don't move out. Stay here and make it better." And I think that's what you have to do wherever you live, whether it's Franklin or Los Angeles. You have to live in the community and you have to contribute to the community. You have to want to make things better. Look at the pond. The fire department built Mitchell Park. It was a hole in the ground. It dressed up that end of the community. And near Hospital Road at the pond, the exempt firemen have started a project to clean up that area. They've put a nice gazebo there with benches and walkways.

I'm working on a bill for a Green Acres program that would include money for historical sites. Organizations will be able to apply for grant money. I've already had a call from the Franklin Historical Society about the bill to see how they can benefit from it. We're founding members, the whole family, of the Franklin Historical Society. There's a lot of history here.

History is extremely important because it tells the citizens that are living in an area now the background of the community and how it got to be what it was. Many people who move to Franklin today have no idea about the mine and its impact. Then they visit the mineral museum and they see what came out of the ground. The Haucks have done a wonderful job with Sterling Hill. They've made it something that people can appreciate when they come and see it. They better understand how hard the work was and how difficult it must have been to work underground and not see any sunlight in rather chilly temperatures, wet and damp, constant dust. People were proud to be sons and daughters of a miner.

I've always loved Franklin and the people who live here. I never had a desire to move anyplace else. Life's all about commitment to a community, and there's a lot of people that are committed to Franklin. Franklin's got a great heart and soul.

One time Franklin miner, successful lawyer, and current bank president Don Kovach is a community member who stayed in Franklin and made it better. Like Senator Littell, Don feels that Franklin does have a viable future. Here Don explains how he started a bank in the facility that his father used for a service station.

My dad became very seriously ill so he couldn't work anymore. That's why the service station had to be closed. A bank came to my dad and we sat down and tried to work out a deal and it didn't work. That made me think that there was no local bank anymore that understood the history of the community, or whether a family name from a character standpoint was worth taking a loan risk, and so forth. So, if this corner was good enough for another bank, I suggested to my dad, "Let's see if I could form a group of businessmen, who would buy his interest out of the real estate, who also then would create a bank by filing for a charter and raising the capital to do that."

That was in 1975. And it happened. We were able to do it. The headquarters are still here. We started out with one employee. Now we have seventy-two. We have branches throughout the county. So we consider our customer base Sussex County, not just Franklin. But I think it's a credit to Franklin that we're here and we're going to stay here. We're community-oriented, there's no question about that.

And, hopefully, we can do more for the community in terms of maybe still rehabilitating the old Main Street area. I've been involved in the Economic Development Commission, and have formed a non-profit corporation, which is dedicated exclusively to redevelopment of the old downtown. That non-profit corporation will be able to raise funds and otherwise buy property, obtain grants and loans, and so forth. It's a non-profit group, but its purpose is to try to foster the rebirth, if you will, of a downtown area. So, if these old storefronts that are now apartments can be reconverted to a store with reasonable rates, I think you'll find — as you have seen in Long Valley, or Chester, or Sugar Loaf, or even Lafayette — that there will be a craft, antique, or other type of business that doesn't necessarily have to be on the highway.

You may even find some professionals that would enjoy that kind of a location. Franklin lends itself to that because the main street is readily accessible from the highway and yet it has an isolation to it that makes it more of a community or old fashioned type miner's village atmosphere. I think it can evolve.

There's a new generation of people who have a great deal of enthusiasm about the past history, and who are talented and want Franklin to be something other than it is at the moment. That history, we can build on that, I think. There's a strain of difference in the Franklin community. Maybe because of our past, but I think they have the grit and the strength that was always recognized by others. Communities in the area and throughout the state recognized that Franklin was different. And that the people were good people and strong. And, I think that if that can filter down to the present day generation — even though they're not natives — then they can take pride in that. And that is really something to go forward on.

James Van Tassel's love for Franklin and his desire to share his enthusiasm with others is what drove him to be active in the Franklin Historical Society. The interview with Jimmy brings the Mining Oral History Project to a close. Here Jimmy comments on Franklin's past, the founding of the society, and the society's Heritage Museum.

It was a nice small town. Everything was simple. You had your Franklin High School and there was a special bonding that the school had because of your football games, your sports, things like that. And there was an old, old bonding in the town itself. While I was growing up, there were still some five-and-dimes left in Franklin on Main Street. We still had the movies. Webb's, which was the newspaper store, was still on Main Street.

There was no Shoprite. None of the stores on Route 23 had even been thought of yet. Main Street was the center of town and the attraction. Everything was up here. We also had the Neighborhood House, which is now torn down. But they used to have canteens and dancing over there.

I had always had an interest in collecting rocks and minerals — some of the minerals being from the mine — and we'd get together with a group of other people who all had the same interest. Franklin had the EDC meetings, Economic Development Committee, and I had gone to a couple of those meetings with a friend. I believe almost two years ago [1995], we all got together and said, "Hey, you know, we can either start some type of society now, or wait until later and say, well gee, why didn't we start it earlier?" So it was either get up and go or just let it lay.

Figure 110. Narrator James Van Tassel is at the far right in this photograph of the officers and trustees of the Franklin Historical Society. (Photo: Courtesy of Franklin Historical Society)

We decided to have a small show up in the old borough hall, last year [1996]. This would coincide with the beginning of the local scouts. They were going to reintroduce the Soap Box Derbies again, like they used to have in the forties and fifties. The local scouts do get all the credit for that because they wanted to bring it back. But we thought it would be nice historically to have something up in the borough hall at the same time as that because they would be generating a crowd down here to watch the show for the kids.

We did it in less than two weeks. At that time about ten people, maybe a couple more were involved. After we had the show, we invited other historical societies over to get the feeling of how to go about this. "All right, what worked for you? What didn't work?" The town was behind us. The mayor and council are still behind us. Before the end of that year, we were incorporated. We formed in September. Had the votes and everything for the officers and trustees, and by December of that year, we had our own incorporation papers.

A lot of the work in getting this building [former Zinc Company time office] was done by Betty Allen. She's a councilwoman. Plus, she's on our board. She had done some work and negotiated through the owners and stuff, and we were able to get this building so that we could occupy it for a small amount. We have gotten excellent display cases through word of mouth. Some have been donated to us. We have purchased some at very reasonable prices. They came from people who were involved with the town, who no longer have stores. Some of the stores were closed and the display cases were left inside. And they, very generously, wanted to donate them or sell them very cheaply.

Our donations came through our members with open eyes. I have a group of members right now, especially our board members and trustees, who knew a lot of people and we have a lot of connections. We keep our eyes open. All completely volunteer. It's all a labor of love. These are all local residents, who had either parents or grandparents who worked for the Zinc Company. This is nice. We're all town people. We're very proud of this building. It's a historical building because it was used by the Zinc Company for personnel and time office. A good majority of the people walked through these turnstiles to go to work.

We want to go hand in hand with the Franklin Mineral Museum and the Ogdensburg mine. Our Heritage Museum is to show how the people lived in Franklin. The ways of life. We can show you stuff from the Franklin High School. Franklin Miners, a semi-professional football team. We're getting artifacts on that. The Franklin Band. We're going to try to get different areas of the town. Also, the police department, ambulance squad, stuff like that so you can see how the people lived in this town.

We have a section out here now of bottles and jugs. We have milk bottles from the local milk farms in Franklin, local bottling companies, that put out different types soda, or beer, or whatever. Old jugs, which probably may have held whisky back during the nineteenth century.

We also have a business section. Almost every building on Main Street, when my parents were growing up, had a store in it. So we're trying to locate relatives and other people that had the stores. We are getting stuff from people. Even if it's just a picture of what the store looked like, or you might have a matchbook cover with the name on it. Everything like that. We're getting all kinds of things for that section, to show people that the stuff was here. It may not be here now, but at one time Franklin really was booming. It had an A & P. It had an Acme. This was right on Main Street. There were butcher stores, five-and-dimes, dentist office, barber shops. You could tell the Zinc Company was right here, and there were two or three hotels.

I'm just so proud of it [the Franklin Historical Society], especially to be the very first president of it, to see it going so well. It hasn't been even a year yet and we have so many goals that we can reach. And, like I said, we have the best bunch of people in the world, the members that we're getting. We're getting more volunteers. We're getting different programs. There's just so much. We're working on getting a tour guide here. Now we're setting up a newsletter.

Sterling Hill has been very cooperative with us as far as helping us get off to a good start, and so has the Franklin Mineral Museum. They're welcoming us, which is great. We're not here to step on your toes or anything. If anything, we want to be like a third part to bring more people here, which is working out so far. When you break it down, if you want to see how people lived, you would come into the Franklin Historical Society [and Heritage Museum]. If you wanted to see the Franklin minerals in general, and the fluorescent room, and stuff like that, then I would suggest that you go to the Franklin Mineral Museum. They have a beautiful display. Now, if you're super interested in mining equipment, the big stuff, and to actually go into the mine and walk around on a tour and stuff, I would suggest that you go to Ogdensburg, because that's a beautiful tour that they take you on. They really know their stuff over there. You could make up like a whole day's tour just to come up and shuttle around the different places.

I've talked to people from different parts of the world over in Ogdensburg because I know the Haucks very well. They're good friends. So I've talked to people from all over the world. Their interest in Franklin is remarkable. European interests in mineral collecting and stuff, Franklin's the place. In geological places, Franklin and Ogdensburg are top notch. You don't realize how famous this town is until you've spoken to people not from this area. We're getting a mixture [of visitors], which surprised me. We're getting both your local and your out-of-town people coming in.

The Franklin Historical Society and the Heritage Museum are intertwined. We meet once a month and everybody's invited to the meetings. We hopefully have a nice future ahead. The future really looks pretty good. We're building a strong foundation here, this first group. If they can keep it going and keep an interest in it within the town itself, I have the feeling that this is going to be here for a long time.

While it is true that no one can predict what the next century will bring to mining communities across America or to the residents of Franklin and Ogdensburg, we can be certain that these communities will look forward to the years ahead with confidence. Most people will agree with Don Kovach. When asked, "What is the best thing about Franklin?" Don proudly replies, "The future."

To those who will be part of that future, may these recollections serve as reminders of the past: the meaningful heritage that belongs to thousands of mining families throughout America, the men who worked in the mines, and the mining companies that shaped a way of life that is no more.

1. Marj Charlier, *Newmont Gold Commemorative Booklet*, 12.

Bibliography

Primary Sources

Baum, John L. Interview by author. Franklin, NJ, 11 September 1995.

Case, Clarence. Interview by author. Ogdensburg, NJ, 7 November 1995.

Cooper, Susan. Interview by author. Ogdensburg, NJ, 4 June 1997.

Davis, Alvah. Interview by author. Hamburg, NJ, 17 October 1996.

Deck, Linda. Interview by author. Washington, DC, 28 June 1996.

Dolan, William C. Interview by author. Ogdensburg, NJ, 18 September 1995.

Falcone, Laura. Interview by author. New Brunswick, NJ, 22 July 1996.

Gerstmann, Ewald. Interview by author. Franklin, NJ, 18 February 1995.

Hadowanetz, Wasco. Interview by author. Ogdensburg, NJ, 12 September 1995.

Hauck, Richard. Interview by author. Ogdensburg, NJ, 6 October 1995.

Hauck, Robert. Interview by author. Ogdensburg, NJ, 6 October 1995.

Kolic, John. Interview by author. Franklin, NJ, 15 May 1996.

Kovach, Donald L. Interview by author. Franklin, NJ, 13 May 1997.

Kozykowski, Bernard. Interview by author. Port Jervis, NY, 18 November 1995.

Littell, Robert E. Interview by author. Franklin, NJ, 24 April 1998.

May, William A. Interview by author. Franklin, NJ, 26 September 1995.

Metsger, Robert W. Interview by author. Sparta, NJ, 11 December 1995.

Mindlin, Edward S. Interview by author. Franklin, NJ, 10 October 1996.

Misiur, Steve. Interview by author. Ogdensburg, NJ, 2 November 1995.

Naisby, John R. Interview by author. Franklin, NJ, 3 June 1996.

Novak, Stephen. Interview by author. Franklin, NJ, 22 September 1995.

Pavia, John L. Interview by author. Ogdensburg, NJ, 13 November 1995.

Revay, Steve. Interview by author. Franklin, NJ, 16 March 1996.

Romaine, Harry. Interview by author. Sparta, NJ, 9 May 1996.

Sabo, Evelyn. Interview by author. Franklin, NJ, 12 May 1995.

Sanford, Steven. Interview by author. Sussex, NJ, 6 May 1996.

Shelton, Robert C. Interview by author. Franklin, NJ, 16 October 1996.

Sliker, Thomas G. Interview by author. Franklin, NJ, 6 May 1996.

Smith, Genevieve. Interview by author. Franklin, NJ, 15 January 1998.

Svecz, Robert. Interview by author. Franklin, NJ, 28 April 1996.

Tillison, Alan. Interview by author. Hamburg, NJ, 4 June 1997.

Trofimuk, Ann. Interview by author. Franklin, NJ, 15 September 1995.

Van Tassel, James. Interview by author. Franklin, NJ, 13 September 1997.

Zipco, Nicholas. Interview by author. Franklin, NJ, 19 October 1995.

Secondary Sources

Books

Allen, James B. *The Company Town in the American West*. Norman, OK: University of Oklahoma Press, 1966.

Alsup, Herbert Justin. *A Brief History of Church Life in Franklin, New Jersey*. Privately printed, n.d.

Aurand, Harold W. *Coal Towns: A Contemporary Perspective, 1899-1923*. Lexington, MA: Ginn Custom Publishing, 1980.

_____. *From the Molly Maguires to the United Mine Workers*. Philadelphia, PA: Temple University Press, 1971.

Barber, John W., and Henry Howe. *Historical Collections of the State of New Jersey*. Newark, NJ: Benjamin Olds Publisher, 1857.

Barendse, Michael A. *Social Expectations and Perception: The Case of the Slavic Anthracite Workers*. University Park, PA: Pennsylvania State University Press, 1981.

Bartoletti, Susan Campbell. *Growing Up in Coal Country*. Boston, MA: Houghton Mifflin, 1996.

Bibliography

Baum, John L. *Three Hundred Years of Mining in Sussex County, New Jersey.* Newton, NJ: Sussex County Historical Society, 1973.

Berthoff, Roland Tappan. *British Immigrants in Industrial America, 1790-1950.* Cambridge, MA: Harvard University Press, 1953.

Bodnar, John. *Workers' World: Kinship, Community and Protest in an Industrial Society, 1900-1940.* Baltimore, MD: Johns Hopkins University Press, 1982.

Brestensky, Dennis, Evelyn A. Hovanec, and Albert N. Skomra. *Patch/Work Voices: The Culture and Lore of a Mining People.* Pittsburgh, PA: University of Pittsburgh Press, 1991.

Brooks, Thomas R. *Toil and Trouble: A History of American Labor.* New York, NY: Delacorte Press, 1964.

Brophy, John. *A Miner's Life.* Madison, WI: University of Wisconsin Press, 1964.

Charlier, Marj. *Newmont Gold Commemorative Booklet.* Denver, CO: The TJFR Group, 1995.

Cook, George H. *Geology of New Jersey.* Newark, NJ: Daily Advertiser Office, 1868.

Cummings, Warren D. *Sussex County: A History.* Newton, NJ: The Rotary Club, 1964.

Chircop, Jeanne, ed. *Facts About Coal.* Washington, DC: National Mining Association, 1999.

Decker, Amelia Stickney. *That Ancient Trail: The Old Mine Road.* Third edition. Trenton, NJ: Trenton Printing Company, 1962.

Dunn, Pete J. *Franklin and Sterling Hill, New Jersey: The world's most magnificent mineral deposits.* Franklin, NJ: Franklin/Ogdensburg Mineralogical Society, 1995.

Garner, John S., ed. *The Company Town: Architecture and Society in the Early Industrial Age.* New York, NY: Oxford University Press, 1992.

Gold Institute. *America's Gold.* Washington, D.C.: The Gold Institute, 1999.

_____. *History of Gold.* Washington, D.C.: The Gold Institute, 1994.

_____. *The Story of Gold.* Washington, D.C.: The Gold Institute, 1997

Goldston, Robert. *The Great Depression.* New York, NY: Fawcett Premier, 1968.

Gordon, Thomas F. *Gazetteer of the State of New Jersey.* Cottonport, LA: Polyanthos Inc., 1973.

Haines, Alanson. A. *Hardyston Memorial.* Newton, NJ: New Jersey Herald Print., 1998.

Handlin, Oscar. *A Pictorial History of Immigration.* New York, NY: Crown, 1972.

Harden, Mildred H., and Ronald H. Keller, eds. *Franklin Borough: Then and Now.* Franklin, NJ: The Franklin Bicentennial Commission, 1976.

Hibbard, Jr., Walter R. *Virginia Coal: An Abridged History.* Blacksburg, VA: Virginia Center for Coal & Energy Research, 1990.

Horuzy, Paul. *The Odyssey of Ogdensburg and the Sterling Zinc Mine.* Ogdensburg, NJ: The Sterling Hill Mining Company, 1990.

Hughes, Florence. *Sociological Work: The New Jersey Zinc Company.* Palmerton, PA: Privately printed, 1914.

Korson, George. *Black Rock: Mining Folklore of the Pennsylvania Dutch.* Baltimore, MD: Johns Hopkins University Press, 1960.

Kushner, Ervan F. *A Guide to Mineral Collecting at Franklin and Sterling Hill, New Jersey.* Paterson, NJ: Ervan F. Kushner Books, 1974.

LaBance, George. *Franklin Borough Golden Jubilee, 1913-1963.* Hamburg, NJ: Wilcox Press, 1963.

LaLone, Mary B., ed. *Appalachian Coal Mining Memories.* Blacksburg, VA: Pocahontas Press, 1997.

Levy, Elizabeth, and Tad Richards. *Struggle and Lose, Struggle and Win: The United Mine Workers.* New York, NY: Four Winds Press, 1977.

Long, Priscilla. *Where the Sun Never Shines.* New York, NY: Paragon House, 1989.

McWhirter, Norris D., ed. *Guinness Book of World Records.* New York, NY: Bantam, 1981.

Morison, Samuel Eliot. *The Oxford History of the American People.* 3 vols. New York, NY: New American Library, 1972.

New Jersey Zinc Company. *The First One Hundred Years of the New Jersey Zinc Company.* New York, NY: Privately printed, 1948.

_____. *Making Zinc: Our Contribution to Better Living.* Franklin, NJ: Privately printed, 1940.

Palache, Charles. *The Minerals of Franklin and Sterling Hill, Sussex County, New Jersey.* US Geological Survey Professional Paper 180.Washington, DC: GPO, 1935. Reprint, Franklin, NJ: Franklin/Ogdensburg Mineralogical Society, 1974.

Ramsey, Robert H. *Men and Mines of Newmont.* New York, NY: Farrar, Straus and Giroux, 1973.

Shepherd, Carol R., ed. *Facts About Minerals.* Washington, DC: National Mining Association, 1999.

Shifflett, Crandall A. *Coal Towns.* Knoxville, TN: The University of Tennessee Press, 1991.

Sloane, Howard N., and Lucille L. Sloane. *A Pictorial History of American Mining.* New York, NY: Crown Publishers, 1970.

Snell, James P. *History of Sussex and Warren Counties.* Middletown, NY: Trumbull Printing, 1971.

Stephens, F. J. *The Preston Letters.* Privately printed, 1976.

Stephens, Matthew. *A Bicentennial History of the First Presbyterian Church of Franklin, New Jersey.* Hamburg, NJ: Wilcox Press, 1975.

Shuster, Elwood Delos. *Historical Notes of the Iron and Zinc Mining Industry in Sussex County, New Jersey.* Franklin, NJ: Kiwanis Club, 1927.

Terkel, Studs. *Division Street: America.* New York, NY: Avon, 1968.

_____. *Hard Times.* New York, NY: Avon, 1971.

_____. *Working.* New York, NY: Avon, 1975.

Thornton, Willis. *Almanac for Americans.* New York, NY: Greenburg Publishers, 1941.

Thurner, Arthur W. *Calumet Copper and People.* Hancock, MI: Privately printed, 1974.

Utah Heritage Foundation. *Celebrate Historic Copperton.* Salt Lake City, UT: Utah Heritage Foundation.1996.

Weiss, Harry B., and Grace M. Weiss. *The Old Copper Mines of New Jersey.* Trenton, NJ: Past Times Press, 1963.

Weiss, Mikki, *et al. Sterling Hill Mining Museum Teachers Guide.* Ogdensburg, NJ: Sterling Hill Mining Museum, 1997.

Wurst, Helen H. *Hardyston Heritage.* Hamburg, NJ: Wilcox Press, 1976.

Wurst, William, Helen H. Wurst, and Richard Haycock. *Souvenir Historical Journal, 200th Anniversary Celebration, Township of Hardyston.* Hamburg, NJ: New Wilcox Press, 1962.

Periodicals

"A Look Back," *New Jersey Herald*, 25 July 1993.

"Fifty-fifth and Final Commencement." Graduation Program of Franklin High School. Hamburg, NJ: Wilcox Press,1982.

"Franklin: Century of Mineral Mining Bonanza," *New Jersey Herald*, 7 March 1998.

"Sterling Hill Makes Improvements," *New Jersey Herald*, 7 March 1998.

"Zinc Miners and Their Leaders," *New Jersey Herald*, 23 May 1946.

Aun, Fred J. "Ogdensbrug Auctions Off Historic Zinc Mine Property for Back Taxes," *The Star Ledger*, 8 March 1989.

Baker, Pamela, ed. *Campaign News, Geology, Gems & Minerals.* vol. 2, Fall/ Winter, Washington, DC: Smithsonian Institution, 1992.

_____. *Campaign News, Geology, Gems & Minerals.* vol. 2, Summer, Washington, DC: Smithsonian Institution, 1993.

Baum, John L. "The Origin of Franklin-Sterling Minerals," *The Picking Table* 12 (February 1971): 6.

_____. "Opening Day Remarks, Franklin Heritage Museum." 29 June 1997.

Franklin Hospital Booklet. Privately Printed, 1908.

Haight, Clarence. "Mining at Franklin and Sterling Hill: Synapse." *The Picking Table* (June 1960): 3.

Mining Engineering. Society for Mining, Metallurgy, and Exploration. Vol. 52, No. 4, Littleton, CO: April 2000.

Office of Development and Public Affairs, "1996 Annual Report," *Quest*, vol. 6, nos. 1 & 2, Washington, DC: Smithsonian Institution, 1997.

Office of Development and Public Affairs, "1997 Annual Report," *Quest*, vol. 7, no. 1. Washington, DC: Smithsonian Institution, 1998.

Point, George. "Digging Through the Past," *Yesterday Today in New Jersey*, June/July 1993.

Sachs, Andrea. "Model Mine," *Asbury Park Press*, 30 November 1994.

Safety Age. New Jersey Zinc Company, May 1939.

Scovern, Peter E., "Miner Symbolizes Former Workers." *New Jersey Herald*, 14 May 1972.

Sterling Hill Mining Museum. Brochure, 1 July 1990.

Women in Mining Education Foundation, "Special 12-page Supplement," n.p., n.d.

Zinc Magazine, New Jersey Zinc Company, New York, NY: October 1924.

Zinc Magazine, New Jersey Zinc Company, New York, NY: December 1954.

Internet Sources

"James Alexander." *Appletons Encyclopedia.* Online. http://www.virtualology.com

"William Alexander." *Appletons Encyclopedia.* Online. http://www.virtualology.com

"Archibald Bruce." *Appletons Encyclopedia.* Online. http://www.virtualology.com

Government Documents

US Bureau of the Census. *Statistical Abstract of the United States.* Eightieth edition. Washington, DC: GPO, 1959.

US Bureau of Mines. *Minerals and YOU.* Washington, DC: Office of Public Information. 1992. Quoted in "Special 12-page Supplement." Distributed by Women In Mining Education Foundation. (n.p., n.d.).

US Bureau of Mines, *This is Mining.* Washington, DC: GPO, 1995.

US Coal Commission. *Report of the US Coal Commission.* Washington, DC: GPO, 1925.

US Department of the Interior. *A Medical Survey of the Bituminous Coal Industry.* Report of the Coal Mines Administration. Washington, DC: GPO, 1947.

US President's Commission on Coal. *The American Coal Miner.* Washington, DC: GPO, 1980.

US Department of Labor. *Historical Summary of Mine Disasters in the United States.* 3 vols., Beaver, WV: National Mine Health and Safety Academy. 1998.

US Department of Labor. *Mine Disasters.* Beaver, WV: National Mine Health and Safety Academy. 2000.

Index

Pratt Institute 127
Presbyterian Church 156, 157, 166, 190, 191, 260, 269
Princeton, NJ 43, 153
Prohibition 225, 227
Pulaski Anthracite Coal Company 234
Purdy, Reginal 246

Rainbow Tunnel 66, 319
Ramsey, Jack 29, 213
Rein, Alan 326
Revay, Steve xi, 30, 42, 83–84, 195, 211–212, 225–228, 235–236, 267–268, 279–280
Richards Mine 277
Riggio, Lou 241
Ringwood iron mines 120
Rock Discovery Center 273, 343, 344, 347
Romaine, Harry xi, 30, 61, 103, 115, 229–230, 271, 318
Rosen, Meyer 39
Rowan, Lieutenant 178
Rutgers University (Rutgers) x, 11, 292, 312, 321
Rutherford Avenue 225, 227

Sabo, Evelyn xi, 30–31, 191–193, 262–263, 278
Sabo, Bill, 213
Safety Age 92, 94, 96
Safety Rules and Regulations 92, 94, 102
Sanford, Stephen xi, 31, 103, 104, 119–121, 124, 141–142, 297–298, 310–311, 317
Scarlet fever 212
Schmidt, Dr. John 207
Sears and Roebuck (Sears) 189, 199, 263

Second World War (World War II, war years) 27, 28, 33, 72, 120, 204, 223, 238, 239, 247, 256, 275, 277, 283, 284
Sentinels of Safety Award 91
Sharpe, Gerald 206
Shelton, Robert C. xi, 16, 31, 86, 166–167, 197–198, 228–229, 282
Shelton, Poo 239
Shuster Park (the park) 169, 232, 261, 264, 269, 271
Siberia 167, 169, 170, 173–176, 210, 295
Sliker, Thomas G. xi, 32, 54–55, 57, 101–103, 116–118, 137, 175–176, 199–201, 222–223, 237– 238, 295–296
Smith, Genevieve xi, 32–33, 168, 206–210, 267
Smithsonian Institution vii, 12, 19–20, 110, 125, 130, 327, 329–337, 340
Snob Hill (Nob Hill) 176, 189
Snyder's Hotel 122
Soap Box Derbies 356
Social Security 155, 156, 162, 170
Sparta, NJ 171, 195, 207, 351
Spenser, Dr. James H. 204–206
SPEX-Gerstmann Collection 309
SPEX Industries 308–309
Stanford University 247
Stankowich, Sashey 239
Star Ledger 287
Stefkowich, Mike 213
Straulina, Frank 226
Sterling Hill 20, 180, 181